Praise for Gary Schwartz

The Sacred Promise

"Gary Schwartz has been on a remarkable journey throughout his lifetime. His body of work is inspirational to me because his passion to bring science and spirit together is dominated by an empirical doctrine that is unimpeachable. His latest research as recounted in this book is likely to impress even the most diehard skeptic. I am proud to recommend [*The Sacred Promise*] and to heap as much praise as possible upon my friend, Gary Schwartz."

—Kelsey Grammer

"For more than two decades I have had the privilege of working with Gary Schwartz. He has amazed me not only with his insights and hard scientific approach, but also his willingness to explore and then share his findings on very controversial topics. Even if you do not believe in the premises of *The Scared Promise*, your intellect and heart will be teased enough so that you just might open your mind and start to believe that all this, and more, is possible."

—Jerry Cohen, CEO, Canyon Ranch

"*The Sacred Promise* is elegant, rigorous, and groundbreaking—research science at its most innovative. Gary Schwartz's exposition of ideas interfacing theory, paradigm, method, and anecdotes shows academia at its best."

—Dr. Lisa Miller, Associate Professor of Psychology, Columbia University, editor of the Oxford University Press *Handbook of Psychology and Spirituality*, and co-host of the TV show *Psychic Kids*

"When I first met Gary Schwartz . . . I was delighted and excited to hear he was researching the question of whether we survive death. This is such an important question, whose answer will determine how we live our lives and our spiritual beliefs. Postmortem survival has long been a taboo topic in scientific circles and Gary had ingenuity and a pioneering spirit to go deeply into this area and not be afraid of what he found. This book shares both his major findings as a scientist and his personal growth and changes during that search. You will be fascinated by this book!"

—Charles T. Tart, PhD, author of *Altered States of Consciousness*, *Transpersonal Psychologies*, and *The End of Materialism*

The Energy Healing Experiments

"Over the past three decades, an impressive body of scientific evidence has quietly emerged that points to stunning new insights into consciousness and its capacity to mediate healing. Dr. Gary Schwartz's *The Energy Healing Experiments* is a stellar introduction to this dazzling evidence, which will transform the very meaning of 'healing'."

—**Larry Dossey**, MD, author of *The Extraordinary Healing Power of Ordinary Things*

"For more than 30 years, I have known and learned from the truly creative genius of Dr. Gary Schwartz. In *The Energy Healing Experiments* his courageous odyssey continues with scientific rigor, penetrating insights, his characteristic twinkling humor, and true wisdom."

—**Kenneth R. Pelletier**, PhD, MD(hc), author of *New Medicine* and international bestseller *Mind As Healer, Mind As Slayer*

The G.O.D. Experiments

"Gary Schwartz has written a provocative book . . . The journey he takes you on is fascinating, mind-opening, and, above all, entertaining."

—**Andrew Weil**, MD, author of the *New York Times* bestseller *Healthy Aging*

The Afterlife Experiments

"This book is an absolute must-read for anyone who struggles with faith, love, death, and aspects of divinity."

—**John Edward**, host of *Crossing Over with John Edward*

"Science meets spiritualism in this extraordinarily precise and detailed chronicle of experiments . . . It is one of the most important books written on this subject."

—**James Van Praagh**, spiritual medium and author of *Talking to Heaven*

"Professor Schwartz exhibits courage and integrity . . . in his groundbreaking experiments. This book . . . is an important milestone in the scientific research on the survival of consciousness after physical death."

—**Richard C. Powell**, vice president for research and graduate studies, University of Arizona

THE SACRED PROMISE

How Science Is Discovering Spirit's Collaboration
with Us in Our Daily Lives

GARY E. SCHWARTZ, PhD

ATRIA BOOKS
New York London Toronto Sydney

 BEYOND WORDS
Hillsboro, Oregon

ATRIA BOOKS
A Division of Simon & Schuster, Inc.
1230 Avenue of the Americas
New York, NY 10020

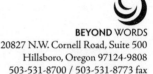

BEYOND WORDS
20827 N.W. Cornell Road, Suite 500
Hillsboro, Oregon 97124-9808
503-531-8700 / 503-531-8773 fax
www.beyondword.com

Managing editor: Lindsay S. Brown
Editor: Julie Steigerwaldt
Copyeditor: Henry Covey
Proofreader: Kristin Thiel
Design: Devon Smith
Composition: William H. Brunson Typography Services

First Atria Books/Beyond Words hardcover edition January 2011

ATRIA BOOKS and colophon are trademarks of Simon & Schuster, Inc.
Beyond Words Publishing is a division of Simon & Schuster, Inc.

For more information about special discounts for bulk purchases,
please contact Simon & Schuster Special Sales at 1-866-506-1949 or
business@simonandschuster.com.

The Simon & Schuster Speakers Bureau can bring authors to your live event.
For more information or to book an event, contact the Simon & Schuster Speakers
Bureau at 1-866-248-3049 or visit our website at www.simonspeakers.com.

Manufactured in the United States of America

10 9 8 7 6 5 4 3 2 1

Library of Congress Cataloging-in-Publication Data

Schwartz, Gary E.
    The sacred promise : how science is discovering spirit's collaboration with us in our daily lives /
Gary E. Schwartz. — 1st Atria Books/Beyond Words hardcover ed.
        p.   cm.
Includes index.
1. Science and spiritualism.   I. Title.
BF1275.S3S341   2011
133.9—dc22
                                                                              2010032485

ISBN: 978-1-58270-258-2
ISBN: 978-1-4391-7742-6 (ebook)

The corporate mission of Beyond Words Publishing, Inc.: *Inspire to Integrity*

For Jerry Cohen,

the Voyager Program,

and "Them Folks"

If a book be false in its facts, disprove them;
if false in its reasoning, refute it. But for God's sake,
let us freely hear both sides if we choose.

—Thomas Jefferson

CONTENTS

CONTENTS

Don't bite my finger; look where I'm pointing.

—Warren S. McCulloch, PhD

FOREWORD

When Dr. Gary Schwartz asked me to write the foreword for this project, I was honored. Our relationship is a professional one that does not include social gatherings or friendly phone chats. We might exchange emails a few times a year to say hello and update each other on our family and work, but I would not have suspected that I would be a candidate to write the foreword for this book. Our history spans ten years and includes four phases of study into the science of the survival of consciousness (which was featured in the HBO documentary *Life Afterlife* and an earlier book he wrote called *The Afterlife Experiments*), multiple verbal debates, and confrontational conversations—all revolving around his professional and personal spiritual evolution or, more appropriately, his spiritual revolution.

FOREWORD

I say *revolution* not only as a play on the word *evolution*, but also because Gary is such a great linguist. His ability to speak to audiences of all ages and walks of life makes him a great communicator and teacher, but it is his childlike exuberance and passion for sharing that knowledge that make his messages so profound.

When you think of the word *scientist*, like most, you might immediately conjure up images of a guy in a laboratory and a white lab coat doing experiments, and it usually has a clinical depiction. You might associate words like *respected*, *accredited*, and *licensed*; but my line of work, mediumship, is usually prefaced with adjectives like *purported*, *self-acclaimed*, and *supposed* instead. I came under a great amount of criticism from colleagues when I told them that I was participating in the studies Gary was heading up.

I was told it was "a trap," or "you are being set up for failure," by a number of people who earn their living from teaching about metaphysics. My answer was always the same: there is a recognized scientist and university that want to give me the opportunity of bringing scientific credibility and understanding to my life's work and passion. How can I not take a stand and participate?

I remember a conversation Gary and I had early on when I said normally people refer to me as either a "freak" or "fraud." Gary's response was, "I would like to see the research change that to 'friend.'" I remember thinking to myself that I did not feel the need to have him as my friend, as nice of a gesture as it was, but the work certainly did. It needed an academic scholar to shed some much-needed light and credibility onto the scene and take it from being a parlor trick, and something that only fools believe in, to a place where it sparks conversations and personal growth, where true soul healing can begin.

I will tell you that a medium comes under a huge level of attack from the skeptics and from those who are so fear-based in their motivations to live the best life possible. Now take someone with real academic credibility, someone who is going to apply stringent double-blind tests to research that deals with the afterlife and the potential to measure it, and that is a bull's-eye on your reputation to attract positivity like no other. I remember being part of a panel discussion on *Larry King Live* one evening and someone asking me why I wouldn't take the million-dollar test that the Amazing Randi was offering. My answer was simple: "Would I want to be tested by someone who has an adjective as their first name or by an accredited professor like Dr. Gary Schwartz . . . like I do?"

After I said that, the level of vitriolic verbal assault on the character of someone who was not even on the panel to defend or explain himself was almost unbelievable. The aspect of Gary Schwartz that I am impressed with the most is his direct communications in the face of criticism from both sides of the belief spectrum. In the face of adversity that is a direct result of challenges to his belief system or lack thereof, he never loses his cool. Instead, he focuses on the mission of raising awareness of the data and how to extrapolate as much of it as possible.

You see, there were many mediums who were expecting Gary to authenticate them because they did a reading as part of the study. Because these were successful, they are now able to say that they are validated. On the opposite end of the spectrum, there were cynics comparing his research to that of photographs of the Loch Ness Monster and Bigfoot and the existence of the tooth fairy. Gary honored his scientific principles, never deviated, continued his research, and kept making the same claim. The claim was that there was so much more here for him, and others, to research and learn about. He wanted to continue to replicate and extend the data and to push the envelope further and further. Only someone who holds to his own truths like Gary does would be able to stay in balance and not be swayed by the strong thoughts and wishes of others.

I remember one of the mediums saying that we were making history and would prove the afterlife. I laughed and said, "*No!* We are just making more work for the scientific community. Once we show them that there is an actual spiritual realm and that we are able to communicate with it under laboratory testing, it will open an abundance of new fields of study for many years to come." I am so pleased to see that projection come true.

The spiritual revolution that Gary has initiated both professionally and in his own life is a journey of enlightenment, with him as our earthly tour guide. He has written a book that is the largest stretch of his scientific academia to date. He has allowed himself to go out on a limb and reach as far as his research mind would allow him, while never losing sight of his humanity. Isn't that the purpose of science and research in the first place, to raise the awareness of the physical world and make it a better place through advancement in thought, technology, and the like?

The Sacred Promise blends within its pages the plight of an academically trained, erudite professor of science and his personal, emotional, and spiritual consciousness, laid bare for all to read. I believe that Gary has done his job well and in the same vein as *The Secret* or *What the "Bleep" Do We Know?!*, *The Sacred Promise* helps us ask more questions of ourselves in the process of reading it.

As a medium, I have to admire the tenacity and passion that Gary brings to his research illuminating lessons on an amorphous topic like grief. His work forces him to score and measure issues that encompass personal tragedies, emotion, and personality. One of his most spoken quotes during the research as a show of support was: "You only need to find one white crow to prove its existence." When I read this book, I was pleased to see the same scientific voice of reason and logic seasoned with the reality that maybe, just maybe, this world of energy that we are all a part of has more layers to it than just what science has been able to teach us.

Thought provoking! Life challenging and life changing! These are phrases that came to mind for me while reading just the first chapter.

FOREWORD

This book forces the readers to examine their own life experiences and question them and take responsibility for their thoughts and actions. It enables you to live a life that is unfolding with assistance from the invisible elephant in the room known as *Spirit*.

Gary Schwartz is a modern-day explorer, sailing off to virtual parts unknown on an ocean of energy that is uncharted for many. His credentials allow safe passage for those who read his research. As our earthly oceans are still teaching us so much, this research promises to do the same for years to come.

As always ... Enjoy the Journey!
John Edward

I'm starting with the man in the mirror.
I'm asking him to change his ways.
And no message could have been any clearer.
If you wanna make the world a better place,
Take a look at yourself,
And then make a change.

—from the song *Man in the Mirror*,
by Siedah Garrett and Glen Ballard

PREFACE

The Urgent Message of the Sacred Promise

Do spirits exist, and if so, can they play a useful if not essential role in our individual and collective lives?

Are there sources of invisible information and guidance waiting to be tapped and harnessed if we are just willing to listen?

If we, in our essence, are spirits too, can we come to see this possibility when we look at ourselves in the mirror? And can we draw on this great inner potential and power with wisdom and love to change our ways accordingly, before it is too late for humankind?

I believe that science not only can address such questions, but in the process can potentially help increase our ability to receive spiritual information accurately, and we can then act upon it safely and wisely.

There Is More Out There

We live in a critical time in human evolution. We are overpopulating the planet and making a mess of the earth in the process. The effects of our unchecked growth are widespread and well documented, from air and water pollution, through the destruction of plant and animal life, to economic and climate instability.

This clear message—that we must change ourselves first to make an impact on the world—is expressed in the megahit "Man in the Mirror," composed by Siedah Garrett and Glen Ballard. Creative artists—be they writers, composers, or painters—often feel that information is given to them from an invisible source. They claim to sense the earth's pain and feel the need to do something about it.

A famous songwriter and spokesperson for personal and global change, as well as a contact with Spirit, was the late John Lennon. He said:

> *When the real music comes to me—the music of the spheres,*
> *the music that surpasseth understanding—that has nothing to do*
> *with me 'cause I'm just the channel. The only joy for me is for it*
> *to be given to me and transcribe it. Like a medium.*
> *Those moments are what I live for.*

Are people like John Lennon brilliant artists but bad psychologists? Are they just weird, if not crazy? Or have they "looked in the mirror" more deeply and honestly than most of us, and have they discovered something fundamental about human nature and spiritual reality in general?

And are deceased people, such as sophisticated scientists like Albert Einstein or compassionate artists like John Lennon, indeed still here, wishing to be of continued assistance to humanity and the planet as a whole? If they and higher spirit beings are showing up in our laboratories and our personal lives, as we explore in this book, then it is only prudent that we try to establish contact.

For if ever there was a time to address the question of a possible Sacred Partnership between us and them, creatively and responsibly, this is it.

Integrating Personal Self-Science with Laboratory Science

The Sacred Promise is about this serious possibility. The book draws on core lessons gleaned from the scientific methods that we can directly apply in our daily lives to potentially connect and communicate with Spirit, to seek help and guidance.

Some of the evidence reported in this book comes from my personal life as a "self-scientist." While these stories speak to the possible reality of Spirit and its potential emerging role in our lives, they are also "proof-of-concept" experiments to allow for the possibility for further research on Spirit's participation in our lives. And they show by example how to apply self-science in your own life, to draw on Spirit's help and guidance.

This self-science is similar to what we all do every day when we follow our hunches and intuitions, or see connections between seemingly random events like repeated references to names or numbers, but it uses the objectivity of the scientific method, i.e., questioning everything and looking for independent confirmation. I believe these are often Spirit's calling cards, and we will explore how to be more watchful for them and how to interpret their appearance.

As part of this exploration, I reveal phenomena that occurred in actual laboratory experiments, intentionally designed and then formally approved by the University of Arizona Human Subjects Committee—as well as exploratory investigations in the early stages of experimental design where we scientists pretested ourselves—that also speak to this emerging reality.

Why Speak Out Now?

Though the critic will be quick to point out that none of this evidence is definitive—and I will be the first to agree—the thoughtful and caring reader will recognize that the information reported in this book speaks strongly to the possibility and promise of Spirit's involvement in our personal lives and Spirit's willingness to help us with day-to-day problems. This is a proof-of-concept book; it provides proof of possibility and promise of our sacred contract with Spirit.

Given that time may be running out for humanity, I feel it is important that we act with some urgency to address this opportunity. I believe—based on the evidence reported in this book—that Spirit is not only willing and able to assist us but is insistent that we proceed and has a broader understanding as to the urgency of the matter. Thus, I have written this book now because I believe, in my heart as well as my head, that if we continue to ignore our true spiritual nature and act in unhealthy ways, we will miss the opportunity to make corrective and wise choices in virtually all aspects of our individual and collective lives.

What Is the Sacred Promise?

The Sacred Promise is all-important. It is more than just the title of this book, and it has wide-reaching implications to understand. There are three levels of meaning to the phrase:

Level I: Spirit has made a sacred commitment to help humanity rediscover its fundamental spiritual nature and to work collaboratively with us to solve our pressing individual and global problems—that is, if we can awaken to Spirit's existence and choose to accept its assistance.

Level II: Spirit has the potential to actively participate in scientific research and be as reliable and responsible as the scientists with whom it is working.

Level III: Science has the potential to determine, beyond a reasonable doubt, that Spirit exists and can play an essential role in our individual and collective lives.

The first level of the Sacred Promise is the tangible commitment that the Spirit world has made to partner with each and every human being to solve our personal and societal problems while guiding us to our own spiritual essence. For many, particularly the scientists, this may seem a grandiose claim. I can assure you it is not. It is not just personal experience but actual scientific evidence—much of which is presented in this book—that demonstrates that Spirit itself is willing, able, and desiring to participate in the evolution toward happiness of all humanity.

For those more cautious in drawing scientific conclusions to what is clearly, from a scientific perspective, a work in progress, there is the second level of meaning to the Sacred Promise, which is easier for the scientist in us to accept. This is the clear indication, from the exploratory investigations and formal experiments reported in the following pages, that Spirit itself is actively participating in the research, which is demonstrating the potential truth of our ultimate hypothesis: that Spirit is the cocreator, with humanity, of humanity's evolution.

For those unable to accept the second hypothesis that Spirit is actively participating as an equal scientific partner in our experiments, on the third level, I hold out this hypothesis: science itself now has the tools to demonstrate that Spirit exists and plays an essential role in our individual and collective lives. For many, this hypothesis alone may have you scratching your head. How can science prove that Spirit exists? Isn't science all about proving that Spirit does not need to exist

PREFACE

to prove our scientific laws and hypotheses? Doesn't the mere acceptance of Spirit interfere with true scientific inquiry? Again the answer is a resounding no.

I may be working in the realm of new science, but my findings are real science and true science. I am not alone in contemplating the existence of a zero point, where all memory and all energy continue to exist forever. For example, Dr. Ervin László has done pioneering work in contemplating the relationship between science and the Akashic Field, and he is but one of many scientists from a variety of disciplines creating experiments to test and verify the role of Spirit in everyday life. The totality of the evidence indicates that science is on the path to proving that a greater spirituality exists.

As we will discuss in Part IV of this book, the word "Spirit" is used generically here and can be likened to the word frequency in physics—it conceptually refers to a wide spectrum of possible spiritual frequencies of energy that potentially can have an effect in the world. It follows that the word "spirits" is more specific and can be thought of as being at least figuratively (and possibly literally) similar to particular bands or patterns of frequencies such as radio waves, light waves, or gamma waves in that they can be clearly distinguished one from another.

The Sacred Promise is intended to open your eyes to the new world before us. As Carole King wrote and sang, "I can see you've had a change in mind; what we need is a change in heart." My hope is that you will read *The Sacred Promise* with both your mind and your heart. There has never been a time of such urgency for hearts to be equal partners with our minds in directing our day-to-day life choices. For me the proposition is straightforward. Each of us needs to answer the question for herself or himself. Are we ready to accept this Sacred Promise to work with Spirit to create a world that honors both our hearts and our minds?

ACKNOWLEDGMENTS

At times our own light goes out and is rekindled by a spark from another person. Each of us has cause to think with deep gratitude of those who have lighted the flame within us.

—Albert Schweitzer, MD

This book is here because of the many people who have lighted the flame within me to write it. Along the way, the light has gone out at times, only to be rekindled by a spark from another person or being, both *here* and apparently from *there* as well. If this book shines, it is because they shine.

There are two people who are particularly responsible for lighting and relighting the flame that produced this book—Susy Smith and Rhonda Eklund-Schwartz. Over the years, I have been blessed to have

ACKNOWLEDGMENTS

known a number of graceful, caring, inspiring, and powerful women—and none exemplify these qualities more than Susy and Rhonda.

My sense is that Susy is insistent that humanity discover that survival of consciousness is as real as the light from distant stars. And Rhonda is adamant, along with her gifted mother, Marcia Eklund, that humanity discover that Spirit is the ultimate reality. Along with my bigger-than-life mother Shirley Schwartz, these women have graciously endured the endless questioning and re-questioning of their agnostic scientist family member in their respective roles as adopted grandmother, spouse, and mother.

If you get the sense that this work has been flamed first and foremost by women, you are correct. I wish to share my inexpressible gratitude for the contributions of a number of women "co-inspirators," women whom you will meet in this book. They include:

Mary Occhino, the star of *Angels on Call*, along with her cadre of angels, who regularly goad me to get past fear and take advantage of the great opportunity to allow Spirit to prove itself. Mary is a true consciousness explorer, as well as a spiritual counselor for people worldwide. For the record, Mary does not favor the word "spirit." She prefers terms such as *Energies* and *Universe*.

Hazel Courteney, whose beautiful books *Divine Interventions, The Evidence for the Sixth Sense*, and *Countdown to Coherence* continue to baffle and inspire me. Some of the most moving and transforming conversations I have experienced about the connections between science and spirit have been with Hazel.

Princess Diana, whose apparent continued association with Hazel, as well as with a number of research mediums, continues to surprise and enlighten me.

Carrie Kennedy, the corporate program director at Canyon Ranch, for her unfailing support and encouragement and for the mani - festations of our synchronous connections that continually remind us that much more is going on here than meets the eye.

ACKNOWLEDGMENTS

Clarissa Siebern, the program coordinator of the Laboratory for Advances in Consciousness and Health, for her dedication and devotion to this work and her courage in sharing her personal and professional journeys in developing her own intuitive skills in communicating with Spirit.

There are many other women, given pseudonyms or not discussed directly in this book, who have played a critical role in this work, and I cannot thank them enough for their contributions. They include mediums, healers, research colleagues, and students. I have intentionally not mentioned some of my research colleagues by name because the focus of this book is so controversial, and there are sometimes "guilt by association" consequences. For the record, this work is not *my* work; it is *our* work, and that includes them and their extended spiritual families.

Besides these amazing women, there have been a number of men who have made essential contributions to this work.

Number 1 is Jerry Cohen, the CEO of Canyon Ranch, to whom this book is dedicated. Jerry has been a longtime behind-the-scenes supporter of the potential existence of Spirit and a larger spiritual reality. Jerry is one of the smartest, wisest, and silliest people I have ever met. Jerry, along with Mel and Enid Zuckermen, has slowly but surely nurtured energy and spirituality at Canyon Ranch. And with Gary Frost, PhD, and Richard Carmona, MD, he has championed responsible and visionary research and applications. Jerry birthed and oversees the Voyager Research Program. Much of the exploratory proof in principle research reported in this book, as well as the existence of this book as a whole, is here because of Jerry. And Jerry appreciates L.O.V.E.

Other men who have been co-inspirators in this work and deserve my deepest appreciation are:

Mark Boccuzzi, my research specialist, whose skills with computers and technology are matched by his compassion for animals and

ACKNOWLEDGMENTS

his commitment to the pursuit of truth, especially regarding the energetic and spiritual mechanisms of mind and health.

Robert Stek, PhD, a "retired" clinical psychologist who waited thirty years to fulfill his dream of doing this type of research. Bob joined the laboratory as a volunteer research associate and has become my colleague and buddy in this work—even if I had the money, his participation is priceless.

John Edward, who sat in the laboratory's "hot seat" for three experiments before he became a superstar medium and who continues to be a champion for reuniting science and Spirit as he educates the world about the reality of mediumship. I am honored by your foreword.

Jonathan Ellerby, PhD, spiritual program director at Canyon Ranch, my younger spiritual brother, who published his first book, *Return to the Sacred*, as this book was being written, and who made me an honorary uncle to his "little Buddha" Narayan, helps to remind me that this work is especially for the children.

Albert Einstein, my physicist and ethicist hero, whose statues and busts sit in both my university and home offices, who persists in attempting to prove that he is still here and continues his passion and commitment for science and world peace. Now that this book is completed, I wonder if it is time for Albert to become formally involved with our research [smile].

There are many other men, not discussed directly in this book or given anonymous names, who have played a critical role in this work, and I cannot thank them enough for their contributions. They include mediums, healers, research colleagues, and students. You know who you are. I am blessed that they are in my life and part of the work.

I wish to highlight some very special people who contributed greatly to the manifestation of this book:

William Gladstone, my agent, adviser, and friend, who wrote the inspirational novel *The Twelve*. This book is here because he is here.

ACKNOWLEDGMENTS

John Nelson, the gifted editor of *The Sacred Promise*, who is a skilled writer of nonfiction and fiction. In numerous phone conversations, John has called me "bro," and I experience this slang title with great reverence (even though I now know that it is used widely by Hawaiians); and Cynthia Black, president and editor in chief of Beyond Words Publishing, and her partner and publisher, Richard Cohn, who not only saw the potential of this book (along with their colleagues, who transformed the rough draft into a polished and beautiful book) but insisted that it be edited by John.

Over the years, the investigations reported in this book have been funded entirely through private gifts or businesses. No federal or state funds have supported this work. Countless people have graciously donated their time and expertise to the work, and I know we all consider it a great privilege to be able to do so.

Other individuals have donated funds—even after they died. Susy Smith, for example, provided funds from her small estate to help the laboratory, and a trickle of funds from her continuing book sales somehow finds its way into her account at the University of Arizona Foundation. To protect donors from unwanted solicitations, I have not included their names in this book. They know already how precious their giving has been. And I know that if a larger spiritual reality exists, their gifts for this work will keep on giving.

Some of the research reported in this book was conducted at the University of Arizona; much of it was privately conducted in my and others' personal lives. Though I have had a few vocal detractors at the university (and elsewhere) who have attempted to discredit me and the work, on the whole, the university administrators, faculty, and students have been for the most part understanding (and even encouraging at times).

The University of Arizona, as a major research and teaching institution, appreciates the importance of academic freedom to pursue knowledge rigorously and creatively. I have mentioned specific administrators and colleagues in previous books, and my gratitude for them grows as my body of work unfolds. I especially wish to thank Dr. Al Kaszniak, professor and former head of the department of psychology, for his wisdom, commitment to integrity, and exceptional humor.

I would like to express my special appreciation and admiration for Allan Hamilton, MD, professor of surgery and psychology and former chair of the department of surgery at the University of Arizona. Allan is a scientist's scientist. His book *The Scalpel and the Soul* is breathtaking, and his work has lighted the flame in me. He has been an ardent supporter of pursing research on the possible roles of energy and Spirit in healing and life enhancement. Though he may not know this, Allan has helped rekindle my passion from time to time.

Finally, I must acknowledge the apparent collaboration of the larger spiritual reality in this book. Scientists can't help using judicious words such as *alleged, apparent,* and *possible*—it's in our blood to be cautious, even when we're being bold. However, let's be blunt for a moment. Some things are ultimately either yes or no:

> Either Spirit exists, or it does not.
> Either Spirit is here, helping to direct this work, or it is not.
> Either Spirit is calling upon us to wake up, to discover our true identity and reality, and to join with it to heal, grow, and transform, or it is not.

If the answers to these questions happen to be no, so be it. We will hopefully learn to live our lives in harmony with nature and each other whether or not a larger spiritual reality exists.

However, if the answer is yes, and we can establish this scientifically, then the Universe is more marvelous and exciting—and filled with more wonders and opportunities—than most of us can imagine.

ACKNOWLEDGMENTS

The history of science reminds us that we have the power of possibility to make these kinds discoveries and transform our minds and hearts in the process. As Einstein said, "One may say the eternal mystery of the world is its comprehensibility."

Many peoples and cultures around the world have believed, and continue to believe, that Spirit is real and plays a fundamental role in our lives and the life of the planet. What science may be doing is finally addressing and validating these experiences and beliefs and hopefully in the process advancing our knowledge and practical applications.

A case in point is the discovery and evolution of flight. The capacity to fly was once only a dream, shared by ancient peoples and evolving cultures as well. But the Wright brothers, and a number of other pioneers, ultimately discovered that we have the potential to create machines that could enable us to fly. The Wrights' first flight lasted only twelve seconds.

Most of us could not have imagined then that in a single century we would have flying machines of all shapes and sizes taking off and landing literally every second of every day around the world. And what is even more amazing to me is what was once previously unimaginable is now taken for granted.

If this book is correct—and I again underscore if—the Spirit airplane appears to be taking flight. Though the metaphorical flight was brief, it looks like it has taken to the air nonetheless. In the process, we are experiencing a Wright brothers moment.

If we too are ultimately spirits, and we have the potential to soar, shall we learn to fly?

If we invest the time and effort to do the science, will our potential connection with Spirit evolve into something as spectacular and reliable as contemporary earth and space flight?

And will we, in the foreseeable future, take all this for granted, too?

If Spirit indeed exists and is calling us, then the truth is that we owe "Them Folks" (as Jerry Cohen affectionately calls them) a tremendous debt of gratitude. They deserve our deepest respect and thankfulness.

Despite my seemingly never-ending agnostic questioning and doubting, in light of the evidence revealed in this book, it seems prudent to hedge my bets at this point and give the spirits the benefit of the doubt.

With the greatest humility and appreciation for your apparent patience, persistence, and participation, if you are really here, we thank you!

PS. I invite you to read the appendices, especially Appendix A, which include key questions and commentary—from self-science to risks in connecting with spirits and unhealthy skepticism, to the opportunity for healing the planet.

PART I

The Sacred Promise

INTRODUCTION

Can Spirit Assist Us with Our Daily Problems?

*Every great advance in science
has issued from a new audacity of imagination.*

—John Dewey

In ancient times, people across the globe not only assumed that Spirit existed, but they fervently believed that their ancestors continued to play a central role in their personal lives and the life of the community. People were encouraged to connect with not only their deceased ancestors and the Great Spirit but also the living Spirit of the earth, animals, plants, and even the stars.

However, humanity gradually shifted from encouraging people to have direct connections with Spirit to sanctioning indirect contacts with a higher spiritual reality via rabbis or priests or other human leaders. In the process, daily spiritual contact was transformed into a practice primarily reserved for temple or church worship and even restricted to weekends.

Today, along with the increased separation of church and state, the idea of connecting directly with Spirit is viewed more as a myth or superstition, or an expression of the misguided experiences of New Age flakes.

I was raised in an agnostic home and was educated in the highly atheistic environment of mainstream Western science. If anyone was encouraged to dismiss the existence of Spirit and its potentially helpful role in human life, it was me. And yet, as I witnessed the extension of human life by the conquest of bacteria, I've also seen chronic lifestyle diseases like arteriosclerosis and diabetes replacing those infections. I began to wonder if our disconnection from Spirit was associated with the chronic emotional and spiritual problems plaguing humankind today.

For instance, one of the greatest public health crises facing humanity, especially people in the West, is excessive weight and obesity. There are numerous apparent reasons for this condition:

1. The easy availability of tasty, high-fat, and high-caloric foods and beverages

2. The massive advertising of fast food restaurants

3. The relative lack of exercise fostered by watching television, surfing the net, and playing extraordinarily realistic computer games

4. The stress and accompanying depression that we experience living in times of serious economic and environmental uncertainty

INTRODUCTION

However, what if all of these causes, as great as they seem, are actually symptoms of an even greater, more fundamental and pervasive cause: our societal separation from Spirit?

What if our increasing feelings of emptiness, loneliness, hopelessness, and meaninglessness are fostered by our belief in a Spirit-less Universe?

What if our physical hunger is actually a symptom of far greater spiritual hunger?

What if Spirit is actually all around us, ready to fill us with energy, hope, and direction if we are ready to cooperate with it?

What if Spiritual Energy is like air and water, readily available for us to draw within if we choose to seek it?

As a prelude to exploring how science is discovering not only that Spirit exists but also that Spirit can collaborate with us in our daily lives, let us consider a real-life situation in which Spirit appears to have played a surprising collaborative role. And as you read this story, I ask that you recall a similar crisis in your own life, or imagine what may have happened to you if you were aware of this resource.

What I find most curious about this story is that it happened to me: a scientist who is open to but also skeptical of such occurrences. Though a definitive scientific explanation for its effect awaits future research, I can attest to the fact that what you are about to read is not misperception or deception.

If we are open to the existence of Spirit and invite its active collaboration, can it actually help us with our daily problems? Since this is the theme of *The Sacred Promise*, I have decided to present this self-science story upfront, without a net, so to speak, or before I've laid the foundation for my case of Spirit's involvement in our personal lives. I would ask the reader to respond with your heart; the rest of the book will present the case for your head.

INTRODUCTION

A Spirit-Assisted Dental Healing

It was October 2008. I had completed most of the research reported in this book. I knew that contemporary science was not only verifying the reality of Spirit but it was also discovering that Spirit could potentially play a powerful role in our personal lives. And for a three-month period, I had been suffering from an increasingly severe gum infection, particularly involving two teeth in the upper right back of my mouth.

In the mid-1980s, an oral surgeon removed a large molar from my mouth. I was told that because the resulting hole was so large—it felt like a cavern to me—I could expect the adjacent two teeth to weaken and fall out within a few years.

Despite his prediction, twenty-five years later I still had those two teeth. However, in the late summer of 2008, the gums around those teeth had become seriously inflamed. The teeth had also become supersensitive to cold and hot substances. I could not put any pressure on them without experiencing severe pain.

Months earlier, my dentist told me that I needed to see a peri-odontist ASAP to avoid infection and that I would probably lose those teeth. I had intuitively felt that I should not seek further dental evaluations and invasive interventions at that time, and later it looked like this decision had been a mistake. (Please note: I am not recommending that anyone avoid obtaining responsible medical or dental advice, nor am I proposing that spirit-assisted healing should be a replacement for conventional medical or dental care.)

It had reached the point where I could drink only lukewarm liquids, and I was restricted to chewing entirely on my left side. Though I had been trained in various energy healing techniques while doing research for an earlier book, and had even begun giving a weekly workshop at Canyon Ranch on basic energy self-care techniques, I had not thought to try either energy healing or spirit-assisted healing on myself.

INTRODUCTION

One night, as my mouth was throbbing and I was musing about all these implications, a novel thought popped into my head. I realized that although we were surrounded by all this air, it was up to us to breathe more deeply if we were to optimize its vitalizing and health effects, and that the same applied to water and that it was up to us to drink more to optimize its vitalizing effects. I wondered if the same applied to all this Spirit or Spiritual Energy around us, and that it was up to us to choose to use its vitalizing health effects.

Simply stated, I realized that Spirit might be like air and water. We can't live without these substances. We typically partake of them unconsciously and naturally. Yet, if we choose, we can learn to more effectively and mindfully partake of them for the sake of our vitality and health.

I wondered what would happen if I asked my purported Spirit helpers—including deceased people and angels—to assist me with my teeth and gum infection. In the same way that we must regularly breathe in air as well as ingest water every day, I wondered whether we needed to create a daily practice of inviting Spirit to assist us. And moreover, I wondered whether we should be mindful when we invite spirits to assist us, and intentionally and consciously collaborate with them.

I wondered if they worked with me in a Sacred Partnership on my teeth and gums, might we achieve a significant positive result together? Of course this was not a formal laboratory experiment; it was a personal exploration—an informal or anecdotal case study.

I was not attempting to confirm, for example, with the aid of a credible and experienced research medium, whether Spirit was showing up when I invited them to help me heal my gums, and I was not being monitored by biomedical equipment. What I was doing was lying in bed late at night, silently attempting to try something I had never done.

I then said to myself something that I hoped I would not regret; I made a specific promise to the Universe. I said that if my teeth and

7

gums showed a rapid, dramatic, and permanent healing—even if they required my daily mindful attention (just as breathing extra air and drinking enough water does), as well as the daily invitation for assistance of Spirit—I would take this healing very seriously. I would not only respect and honor it; I would make sure that others might benefit from it as well.

To my utter amazement, over the course of the next few days, my inflamed gums returned to a healthier state. My gums stopped bleeding when I brushed them, and they were minimally sensitive to touch. The persistent heat and cold sensitivity, as well as the pain, which had been present for months, decreased dramatically.

I then extended the spirit-assisted dental practice and healing to the whole of my mouth. At the time I edited this chapter, in August 2010, my entire mouth was in the best shape it had been in for more than four years. I did not have the courage to return to my dentist and try to explain what may have occurred. Though he seemed like an open-minded healthcare provider, the idea of spirit-assisted dental healing might be too much for him to swallow.

Of course, we could speculate that my teeth and gums might have healed on their own, without any assistance from them or me. It's called spontaneous remission. However, this was not my dentist's prognosis based on more than twenty years of clinical practice. His prediction was that my gums and teeth would get worse, not better.

We could also speculate that the healing was entirely the result of a mind-body effect. Maybe because I believed that my dental problems would be healed, my teeth and gums followed my expectations—the placebo effect. Maybe no Spirit was involved.

However, the truth is that I did not know if my teeth and gums would heal. I had never read or heard of cases using one's mind to treat severe gum and tooth disease. Such cases may exist, but I did not know of them. For me, this was a personal experiment in the sense that my attitude was, "I don't know. Could be yes, could be no;

show me the data, I'm open." This is the essence of a proof-of-concept experiment: ascertaining that an effect happened and that research is needed to determine its true cause.

And of course, I do not know what will happen to my gums and teeth in the future. All I know is what has transpired over the past two years. My repeated experience has been that when I remember to invite Spirit to assist with my teeth, just as if I remember to take deep breaths or drink more water, my health increases accordingly.

By itself, my positive dental healing experience means nothing scientifically—though my mouth is very happy, regardless of the actual mechanisms involved. However, in light of the many controlled research experiments as well as exploratory observations you will read in this book, my positive dental healing is potentially noteworthy. It illustrates the possibility that Spirit assistance may play a meaningful role in dental as well as general healthcare.

Of course, only you live in your body and me in mine. Only you can make the choice of breathing more deeply, drinking more water, eating less food, drinking less alcohol, and inviting Spirit to be an integral part of your life. The same applies to me.

It is not enough simply to go to a house of worship, pray for help, and then assume that all will be handled. Being religious per se does not guarantee that we will be healthy or happy. What I am speaking of here is a sacred partnership, a true collaboration between Spirit and us. It is not an "either/or" relationship; it is an "and" relationship, an active partnership.

Many of us, consciously or unconsciously, assume there is no Spirit to assist us in our lives. Because of this self-imposed spiritual void, we may be prone to overeat, drink alcohol, and take more pharmaceutical drugs as a consequence.

But what if we are wrong? Then we are making a serious mistake.

If we live in a Spirit-full Universe instead of a Spiritless Universe, and if many of our greatest problems directly or indirectly stem from our false belief in a Spirit-less Universe, then it seems high time that we reexamine our potentially erroneous belief and do the necessary science to get a clear answer, one way or the other.

What do you think? Can you imagine asking Spirit to assist you with your health or economic crisis right now, or to heal the break with your mother, or to protect your child in the military? Think about it. Just asking may not be enough; you may have to believe and be open to it, and for most of us that means "show me," which is why I'm writing this book.

As you take this journey of professional and personal scientific discovery with me, you may find that your thinking about this sacred partnership will be as radically and permanently transformed by the emerging evidence as mine was.

The journey awaits you.

1

THE GRAVITY OF SPIRIT

The important thing is not to stop asking questions.

—Albert Einstein

Consider the following questions:

Does our consciousness, including our personalities and memories, survive physical death?

Do each of us have personal spirit guides, sometimes called guardian angels, and can they play an active role in guiding our lives?

Can Spirit play a fundamental role in healing and health?

Can we literally learn to call on Spirit, which includes our deceased loved ones, higher spirit guides, and the Great Spirit or the Sacred, for healing ourselves and reclaiming the planet as a whole?

I am going to ask you to imagine what it would mean to you, and the world as a whole, if we could establish scientifically—once and for all—that yes is indeed the correct and accurate answer to these questions.

In this book, you will read for the first time about ongoing research in my laboratory that points inexorably to the emerging conclusion that all these possibilities and more are real. And, even more astonishing, I will show you how these possibilities materialize in my own life and those around me, and how you can learn to spot them in your life and begin to establish your own sacred partnership.

Einstein said, "Imagination is more important than knowledge." So, let's imagine, for the sake of argument, that there really is a larger spiritual reality.

Let's imagine, for the moment, that we are, in our essence, spiritual beings having a physical experience.

Let's imagine that what is preventing our ability to connect with this larger spiritual reality is our relative lack of knowledge and maturity as a species and our disconnection from nature, which is often the bridge between the two.

Let's imagine that, in the same way we used to believe that the earth was flat, that the sun revolved around the earth, that objects were solid, and that the so-called vacuum was empty—and these beliefs were all discovered to be fundamentally wrong—that our current Western scientific beliefs about a nonintelligent and Spiritless Universe are also fundamentally wrong.

Let's imagine that just as we did not know until relatively recently that invisible fields of energy could carry extraordinarily complex patterns of parallel information, making it possible for billions of us to be globally interconnected by cell phones and even be located by GPS devices, that these same invisible fields of energy also carry complex patterns of information associated with a larger spiritual reality.

Let's imagine that it is only a matter of time before we discover the physics and invent the technology to evolve from the cell phone to the "soul phone"—an advanced technology that will enable us to communicate accurately with this larger spiritual reality.

If you have difficulty imagining these possibilities, remember that there was a time when people could not imagine supersonic airplanes and spaceships, global satellite communication systems, high-definition digital televisions, or even pocket-sized gigahertz computerized smart phones.

So, let's again imagine that with the aid of this advanced spiritual technology, we will have access to information and guidance that can help us overcome our dangerous conceptual and behavioral programming that perpetuates unhealthy and unwise practices harming ourselves, others, and the planet as a whole—like treating the natural world as dead and lifeless, only to be exploited for our own designs.

Let's imagine that there exists a Sacred and Infinite Intelligence that we can come to increasingly know and live in harmony with its greater unfoldment.

And let's imagine that expanded science, created and employed by the human mind, has the capacity to reveal all this and more.

The Nature of the Research

The fact is that the media, as well as academia and even organized religion, is on the whole super-phobic about these possibilities being provable. They would supplant their authority and rule, which may be helpful in their minds, but is a hindrance to our evolution toward knowing Spirit.

If there is a larger spiritual reality—and I underscore if—then these individuals and institutions are doing our species and our planet a profound disservice, especially in these critical times. They are actively impeding the discovery of a deep truth that could foster our survival and transform our lives, present and future, into a veritable paradise on earth.

If my emerging research is correct—and I again underscore if—then the larger spiritual reality is making a Sacred Promise to assist us in our healing and evolution. That is, if we are ready and willing to listen.

I have written *The Sacred Promise* to inspire you to seriously consider the possibility that science is on the verge of making momentous discoveries concerning the existence and nature of the human soul, higher spiritual beings, and the Source of it all—and how they can be in essential partnership with you. That means helping and guiding you in both the big and small decisions of your life on a daily basis.

This proof-of-concept book is an urgent call to action for us to listen and watch more closely for Spirit's invitation to collaborate with it in our daily lives. I share for the first time how evidence for a larger spiritual reality is showing up not only in my university laboratory but also in the laboratory of my personal life (as related in this book's introduction), and in the lives of others as well. I am revealing this extraordinary research, involving both laboratory-based science as well as self-science, to document how this is not only applicable broadly to the field of science but directly to our personal lives as well. As voiced in my acknowledgments, humanity appears to be having a Wright brothers moment of exceptional significance.

I am revealing this controversial information and my expanded methodology with serious trepidation. I realize that in not only sharing my serious interest in these questions but also delving into compelling personal experiences I breech the boundary between formal scientific research and the laboratory of personal experience, albeit with the same critical mind. As such, some readers will say I've become too close to my subject or too subjective in my approach.

I understand these concerns and questions and regularly ask them of myself (and ask others who work closely with me to question, too). It is responsible as well as sane to do so. However, if what I am researching in this book is real—and for the record, everything I have written about really happened—then the implications are sufficiently profound that we will all have to reexamine some of our most cherished assumptions and beliefs, and as such I can't be fainthearted.

This is unavoidable—it comes with the territory. Reexamining our belief systems becomes our deepest challenge, yet it affords us the greatest opportunity for healing and transformation. This is central to fulfilling *The Sacred Promise*.

The Importance of Personal Self-Science, Exploratory Investigations, and IRB Human Subjects Research

What exactly is research, and how does it relate to self-science? The word *research* has both general and specialized meanings, and it is essential that I clarify how the term is being used in this book.

In everyday language, the word *research* has a broad spectrum of applications. Research literally means "to re-search," i.e., to search closely, again and again. We can do many kinds of research, including library research, field research, personal research, informal research, exploratory research, pilot research, systematic research, highly controlled research, computer modeling research, proof-of-concept research, hypothesis-testing research, and replication research. We can research virtually anything, from subatomic particles to superclusters of galaxies and everything in between. We can research body, mind, and spirit. We all have experimenting minds, and we love to discover new things and understand how things work.

The federal government has established important protective guidelines for different kinds of research, including research with humans and animals. Though this may sound strange, to adhere to federal guidelines, universities typically ask the following question: (By necessity the next few paragraphs are a bit technical—they sound like legalese—so please bear with me. The quote below comes from the University of Arizona website concerning research with human subjects.)

> Is this project a <u>systematic</u> investigation <u>designed</u> to develop or contribute to generalizable <u>knowledge</u> (including use for a thesis or dissertation, publication or poster presentation)? **If NO**, the project is *not* considered research; IRB review is *not* required.
> [Note that the <u>underlining</u>, **bolding**, and CAPS come from the website; I added the *italics*.]

Yes, you read correctly; such a project is *not* considered research. From this specialized perspective, two kinds of investigations reported in this book are not officially considered research. The reasoning is that these investigations are not, by themselves, generalizable—for example, they are often based on a single instance—and they are frequently not systematic. Hence (1) careful observations made in one's personal life and (2) exploratory investigations where university scientists pretest themselves before performing subsequent controlled research with recruited subjects are not considered research in the eyes of the federal government or university IRB (institutional review board).

However, this federally restricted use of the word *research* does not mean that the observations from personal life (herein called Type I Investigations) or experimenter pretesting (Type II Investigations) are either unimportant or uninformative and should therefore be dismissed. Quite the contrary, what we are calling self-science—conducted in the living laboratories of our personal lives (Type I) or exploratory investigations where laboratory scientists pretest themselves (Type II)—sometimes uncovers essential proof-of-concept evidence that provides the foundation for advancing science and society. Moreover, when Type I and Type II information is carefully combined with information obtained from larger scale systematic and intentionally designed generalizable research (Type III Investigations), the whole is greater than the sum of its parts.

To preserve this clarity, and honor the federal guidelines as implemented by the University of Arizona IRB, we refer to Type I and Type II investigations as personal investigations and exploratory investigations, respectively. However, sometimes the word *research* is by necessity employed in the general everyday context of the word because it is the clearest way to convey the fact that the scientific method can be applied in our personal lives as well as in university laboratories. The process of self-science extends beyond restricted semantic uses of the term *research*; it encompasses the deepest sense of what it means to *do* research.

What Manner of Evidence Will Prove the Existence of Spirit?

Although I am philosophically an orthodox agnostic—and maybe because I am—I strongly appreciate the clever statement: "Moderation in everything, including moderation." Thus, it follows that it is prudent to "question everything, including the questioning of everything." In other words, sometimes questioning or the scientific method has its limitations.

Every now and again science discovers that something is reliable and true, even if the scientists do not know precisely how it works or why. Let's consider the case of gravity. You experience this every day when you step out of a vehicle or walk down a flight of stairs. Physicists and engineers have made tremendous progress in studying the effects of gravity, to the point where we can launch and position stationary satellites in space and even land robotic probes on the moon, Mars, and places beyond. However, it is important to note that, while we can achieve these incredible feats of technology, we still do not know for sure whether gravity is a physical force, the bending of space-time, or a property of superstrings that actually creates mass (rather than the other way around); and there are additional credible theories about gravity still in contention. The analogy here is realizing that we can investigate whether something exists or does not by its effects alone, even if we are unable, with complete assurance, to come up with an ultimate theory to explain it.

You are about to read a collection of experiments and investigations conducted in the university laboratory (Type III) as well as the laboratory of my personal life (Type I) and that of others—including pretests where the researchers conducted preliminary investigations on themselves (Type II)—which all point inexorably to the existence of Spirit and our collaboration with a larger spiritual reality. None of the experiments or investigations by themselves are definitive. Moreover, many of them are too exploratory, as well as controversial, to be published in mainstream scientific journals (and I have published hundreds of scientific papers in such journals).

However, this does not mean that the findings reported herein lack scientific validity and utility. Quite the contrary, each experiment or investigation is significant and offers meaningful conclusions as it raises important questions for future research and applications. Again, we are at the proof-of-concept stage in this exploration of Spirit and its possible relationship with us. Note that it is the combination of these daring and unorthodox experiments and investigations—and the complementary and synergistic nature of their conclusions—that together establishes a compelling case for science to address these fundamental questions creatively, comprehensively, and definitively.

If ever there was a place and a time for us to remain mindful that the whole is greater than the sum of its parts, it is here and now.

Should science take on the task? I believe it must. The promise of this research is too great. The implications for our personal lives and the planet as a whole are too broad and too deep. And the opportunities afforded by this research for personal and global healing and transformation are too far-reaching.

Yes, there are potential risks in doing this research and connecting with Spirit. I address some of these important questions and concerns at the book's conclusion. However, if the premise of this book is correct, our survival and evolution as a species may well depend upon whether we rise to the occasion and perform this critical research carefully yet boldly.

If the grandest implication of the emerging research in this field is proven to be correct, Spirit is making a sacred promise that it will increasingly help guide us toward collective health and prosperity, toward solving some of our greatest challenges—that is, if we are prepared to listen and act in concert with it.

We can all participate in this research, because the call of Spirit and its applications apply to each and every one of us—presuming of course that we are willing to expand our minds and do the work.

Each of our lives is a living laboratory, and some discoveries can only be made by us as individuals. For instance, if you have a health issue and you ask for Spirit's assistance, and then an hour later a friend recommends an alternative approach that later proves fruitful, you could call it a coincidence or even a synchronicity. But if this scenario is continually repeated, your life has given you the best proof of Spirit possible: real effects in the real world.

This kind of spirit-assisted guidance and healing points to the phenomenon of what Dr. Carl Jung called synchronicity as a measuring stick. Briefly, synchronicity is when two or more events seem connected and are subject to a similar influence or energy but lack a cause-and-effect relationship. The fact is, only you can observe what happens within and around you. Only you can discover the extent to which you can develop an effective partnership with Spirit that improves your life in practical ways. Only you can discover patterns of events that cluster together in meaningful ways and provide encouragement and guidance in the progress of your life.

Besides addressing this great purpose, I have written this book as a true-life spiritual adventure. As you will see, this journey to reunite science and spirit—and reconnect Spirit to each of us—has been full of wonder, a tasty treat for curious minds and caring hearts, and great fun, if nothing else.

And the journey is just beginning . . .

2

PARTNERSHIP WITH SPIRIT

Everyone is entitled to their own opinion, but not their own facts.

—Daniel Patrick Moynihan

If Spirits are real, can they play a fundamental role in your life?

There is growing scientific evidence that this is occurring, but just as crucial is the testimony of people from all walks of life who claim it to be true for them. Over the past ten years, I have come to know, investigate, and ultimately befriend a group of people who are convinced that not only are Spirits real but also that they are in daily contact with these invisible entities or energies. These people insist that they do not just believe but know that Spirits play a fundamental role in their lives. They experience these Spirits as helpers or guides in their lives. For Clarissa Siebern, whom you will meet in chapter 15, her alleged Spirit guides instruct her to say or do novel things that turn out to be timely and meaningful, and are evidential of her emerging intuitive abilities. For example, she was prompted to give me a Salvador Dali painting, *Meditative Rose*, which led to a series of rose synchronicities ending five days later when I met my future wife, Rhonda, the Greek origin of whose name means rose.

These people are convinced that their respective personal Spirit guides—be they deceased people, higher entities including angels, or the Great Spirit itself—are their friends, companions, extended family, and partners. Some hold ongoing two-way conversations with them in their minds, asking for advice and guidance as they would of a trusted friend or wise counselor, while others just put out their requests and look to their lives for the results. For instance, they may do one thing prompted by Spirit, which turns out to be a charitable contribution, but receive the equivalent compensation from another source within days. You may have experienced something similar in your own life but didn't make the connection.

I call this the sacred partnership hypothesis because these people experience the collaboration of Spirit as reflecting a relationship whose intent is global as well as personal. They are convinced that this emerging sacred partnership exists for their personal benefit and for the well-being of humanity and the planet as a whole. They experience the sacredness of this partnership as feelings of love, trust, gratitude, wholeness, reverence, and transcendence.

These people come from all walks of life, and they are highly successful in their respective lives. They include:

- a senior vice president of a major investment company
- an administrative staff person at a major university
- a number of psychiatrists (MDs) and clinical psychologists (PhDs)

+ a computer engineer and senior programmer at a Fortune 100 software company
+ a biophysicist researcher turned a spiritual healer
+ a former mathematics teacher

You will meet some of these people in this book in the context of their contributions to scientific evidence addressing the sacred partnership hypothesis.

My mainstream academic colleagues, as a rule, would presume that such people must be naive, ignorant, foolish, self-deceptive, fraudulent, or even crazy. And since history reminds us that some of the most insane and destructive individuals have believed that they were doing the work of Spirit, if not God, it is easy to understand and sympathize with those who hold such skeptical opinions. For example, the serial killer David Berkowitz, known as the Son of Sam, claimed that he was following instructions from "father Sam," and that he was a victim of demonic possession. We will return to this issue at the end of the book.

However, the spirit-guided people I am including here do not show the classic signs of schizophrenia or other forms of mental illness. They do not fit the stereotype of New Age flakes. They are not loose thinkers, though they often think outside the box. They are not frauds, and they are not actively engaged in self-deceptive or destructive acts.

These people hold responsible jobs and are typically excellent communicators. Some have published books, and others are currently writing them. They show advanced proficiency as well as a preference for integrating reason with intuition in their daily lives. They all show demonstrable evidence of paranormal-like abilities in their daily lives, like the investment banker who repeatedly followed his hunches in recommending emerging technologies that have created windfalls for clients, or the psychiatrist who sees dreamlike images in sessions with clients that often lead to breakthroughs in their therapy.

They are typically kind, caring, creative, ethical, and compassionate people. And they all enjoy a great degree of playfulness and senses of humor—which makes them a great fun to be with.

However, is there any scientific evidence that what these people know to be true is in fact actually real? Do Spirits actually exist, and if they do, can they play a close and useful collaborative role in our lives?

As Daniel Patrick Moynihan, the late New York senator, said, "People are entitled to their own opinions, but not their own facts." How much of what they allegedly know are their opinions, and what actually are the facts? Where's the evidence? Where's Waldo?

To answer such challenging questions, society often turns to science. Contemporary science has provided us with many opinions or judgments about the reality of Spirit that are often critical if not dismissive. However, these negative opinions are based almost entirely on prevailing materialistic theories, in the virtual absence of factual research about Spirit.

This is about to change and the world as we know it will never be the same. As *The Sacred Promise* reveals, cutting-edge science is ready to rediscover Spirit in inventive, inspiring, and enlightening ways.

An Ancient and Contemporary Hypothesis about the Invisible

Since the beginning of recorded history, people have believed in the existence of some sort of a spirit world that played a fundamental collaborative role in the creation, manifestation, and transformation of the physical world. The spirit world was presumed to be nonphysical because it could not be readily perceived by our conventional physical senses of seeing, hearing, tasting, smelling, and touching.

The Spirit world was invisible to most of us most of the time. However, certain individuals—sometimes called shamans, medicine men and women, and seers—under certain circumstances could expe-

rience the unseen world and consciously communicate, if not collaborate, with this supposedly higher or greater reality.

The Mayan shamans, for example, discovered the black hole at the
center of the Milky Way, and their astronomers ascertained the preces-
sion of the equinoxes long before modern science did. And the Dogon
tribe in Mali knew about the existence of Sirius B before twentieth-
century astronomy verified the claim. (Without telescopes the Dogon
people could have only known this fact by some sort of firsthand experience, such as spirit travel, which historical records describe.)

Going back to the 1960s, there has been a resurgence of public
interest in shamanism, spiritual healing, and psychic abilities, including the possibility of foreseeing the future and speaking with the
deceased. The media has capitalized on this public interest, and there
are presently more television and radio shows devoted to these possibilities than ever.

One reason the public is open to the invisible world is because we
now live in an age of the practical invisible. Prior to the twentieth century, these worlds were solely the province of shamans, psychics,
healers, and mystics. However, with the discovery of radio waves and
the evolution of television and satellite communication, this invisible
science has had a great impact on our daily lives.

Today we watch high-definition television provided by electromagnetic waves transmitted from satellites in space. We communicate
globally with smart digital cell phones that contain powerful micro-
computers for text messaging, surfing the internet, receiving online
video broadcasts from the media or our friends, and even playing
sophisticated games.

We have a large collection of remote controls for our television
receivers, DVD players, gaming stations, and surround-sound enter-
tainment systems using invisible microwaves. And some of us even
cook our TV dinners with the same technology—unless we're yogis
living on pure prana.

We live within a pervasive web of wireless communications. And all of this amazing technology is based upon the informational capacity of the invisible world of electromagnetic frequencies. In fact, most of this science is based on the abstract theories of quantum physics, which revealed the microcosmic world hiding within the seemingly solid, observed world of classical science.

It takes relatively little imagination to move from the practical invisible of today's technology to the possibility of the shaman's invisible world as envisioned by ancient peoples. Maybe our cell phones can even be adapted for that use, like the Peruvian Whistling Vessels. These clay pots were thought to be water jugs for centuries, until someone blew into their long spout, which gave off a tonal frequency that alters one's consciousness. Some speculate this was the frequency technology the shamans used for journeying to other worlds.

Why Mainstream Science Assumes that the Spirit World Must Be Nonexistent

When I was pursuing my PhD in psychology at Harvard University in the late 1960s, there was a revolution in psychology taking place. Distinguished experimental psychologists who followed the Skinnerian behavioral framework—after Professor B. F. Skinner from Harvard—claimed that our thoughts and feelings were either illusions or immaterial to how humans functioned.

Subjective experience, especially spiritual, was presumed at that time to be an epiphenomenon—a secondary phenomenon of little importance. In medicine, according to Webster's dictionary, an epiphenomenon is "an accidental or accessory event in the course of a disease but not necessarily related to that disease." The M and C words (*mind* and *consciousness*) were considered to be taboo in the medical world's lexicon.

We now know that this then-popular scientific view was nearsighted, to put it mildly. However, today the S words (for *spirit, soul, sacred*) are still considered to be taboo in much of science. Belief in Spirit is akin to superstition or stupidity. Spirituality is often presumed to be an epiphenomenon or side effect of the brain and our need for hope, even if the hope is delusional.

Why is mainstream science typically so dismissive of the possibility of Spirit and why should we care?

As mentioned briefly above, this reflects the sordid history of fraud and pathology associated with a subset of self-proclaimed spiritual people, including highly visible religious and political leaders, as well as self-proclaimed psychics and healers. However, there other reasons why mainstream science dismisses the existence of Spirit, and they include the following:

+ Historically genuine scientific findings were sometimes dismissed or suppressed by powerful religious institutions. In the Middle Ages, scientists were threatened and even jailed for challenging prevailing Church doctrine.
+ A corollary of this suppression was that science as a discipline needed to divorce itself from anything religious or spiritual in order to freely seek empirical truth and develop methods to discern fact from faith, fancy, and fiction.
+ In the process, scientists began to shift their focus from pure science and truth-seeking to "scientism" that creates its own dogma. Thus, science became equated with a material, nonspiritual theory of the universe. When the foundation of science shifted from the process of discovery to theory/philosophy, the possibility of discovering nonphysical aspects of nature was minimized, if not suppressed.

However, when mainstream science as an institution focuses on opinions at the expense of facts, it can dismiss or ignore some of the profound challenges facing society and the planet at this moment in history, and this can greatly impact our lives. For example, for eight years the Bush administration sought out those scientists who disputed the clear evidence of global warming and used this "science" to ignore the problem. Science should be about evidence, not opinion.

Consider this: If you were Spirit, and you were invited to participate in research with scientists on the earth, would you prefer to work with people who are kind and playful and who genuinely care about discovering the truth, or would you choose to work with people who are hostile and care more about proving or disproving specific theories than discovering something that might be different from what they believed? Whose experiments would attract your participation? I sometimes wonder if one of the reasons why my colleagues and I are fortunate in often obtaining positive results is because Spirit knows that we really care about finding the truth of the matter, and that we will follow the data where it leads.

This book focuses on the emerging reality of Spirit—rediscovered using the scientific method as expanded by self-science—and the promise of Spirit's increasing helpful role in our lives. However, its subtext addresses the nature of science versus scientism in the larger context of our evolving understanding of a greater spiritual reality and living in harmony with both the physical and the spiritual.

What Do You Believe about the Existence of Spirit?

How would you answer the following questions, using a −3 to +3 rating system:

+3 Definitely YES   −1 Possibly NO
+2 Probably YES   −2 Probably NO
+1 Possibly YES   −3 Definitely NO
 0 Do not know one way or the other

Does Spirit exist and play a fundamental role in our lives?

Could the synchronicities of daily life be Spirit directed?
Can we rely on our Spirit connection for guidance?

People's opinions can range from superskeptical disbelief (a firm −3 NO) to fully formed faith (a +3 YES), with representation at every level in between on such questions. Your response here and elsewhere—whether directly asked or not—is an important barometer to test your openness or resistance to any area of inquiry. Ultimately, it is up to you as an individual to make your own leap of faith or feeling. But first, you must know precisely where you stand.

When I began this research over a decade ago, I was somewhere between a 0 and a +1—maybe around a +0.3 or +0.4. In other words, I was somewhat open to the possibility that Spirit might be real and could potentially play some sort of role in our lives. Notice my use of the terms "somewhat open" coupled with the words "might be," "could potentially play," and "some sort of role." I was firmly somewhere between a 0 and a 1. This was partly due to my questioning mind and partly to my understanding of quantum physics, which leaves the door slightly ajar for such a possibility.

I happened to have been raised in a strict reform Jewish home, which presumed "ashes to ashes, dust to dust, case closed." And my scientific education in psychology and physiology professed that the mind was entirely a byproduct of the brain. Therefore, because of my background, any experience of Spirit would be classified as superstition or stupidity—if not a sign of self-deception or psychosis. Based upon my upbringing and education, I had good reason to disbelieve in the possibility of Spirit, and I knew how to adopt a −3 position. FYI, I still do. It's like riding a bike; once you learn how, you never forget.

As you read this book, you will come to discover how it is that I slowly but surely came to change my rating from +0.3/+0.4 to +2.8/+2.9, and it is my hope that the same happens to you. When you carefully examine the totality of the evidence reported in this book, and you consider the possibility that these findings are just the tip of the iceberg, you may come to agree with me that it is only die-hard obstinacy that motivates dedicated disbelievers.

Again what is fascinating about the evidence reported in this book is that it combines exploratory investigations (Type II) and systematic experiments (Type III) conducted in scientific laboratories with data collected in the natural laboratory of our daily lives (Type I). This is self-science, and this type of data should be considered objectively. Carefully documented evidence from people's personal lives compliments what the laboratory research is uncovering— whether the laboratory research is physically conducted within a formal academic setting or in one's home. I think that the combination of both kinds of evidence (informal/personal and formal/laboratory) proves to be more compelling and convincing than either one alone; together they establish convincing proof of concept.

It is one thing to observe a phenomenon in a basic science laboratory; it is another to observe it in our daily lives. It is one thing to have a medium confirm facts about the deceased in a double-blind experiment; it's another to hear a voice saying or to get an intuition to pull off the freeway and ride out a thunderstorm that creates a pileup two miles down the road. When the types of evidence converge, and do so repeatedly, we not only have more reason to believe in the reality of the phenomenon, but we also have reason to incorporate this knowledge into our personal lives and act upon it.

Three Categories of Evidence and the Journey of Discovery

In addition to employing the three types of investigations, we will be examining three categories of evidence in this book. You will be better able to take the journey of discovery with me if you can appreciate how I reached my conclusions—and in the process, you can come to your own conclusions—and you can share in the excitement, wonder, frustration, confusion, and fun of actually doing the work with me.

Again, as Carole King reminds us, this is a journey from the head to the heart at every stage, and while we may occasionally make quantum leaps, rigorous examination is a good starting point. As we delve into each category, I ask that you think about your own history and how this exploration may explain events that until now were unexplainable or even those that were ignored at the time but now come to mind.

The first category of evidence addresses the possibility that people's consciousness survives physical death and that their consciousness is as alive—and is as willful/intentional—as yours and mine. You will see how I came to conclude that information provided in investigations and experiments with research mediums indicates that the spirits appear to not only be alive and well, but as intelligent, willful, and playful as we are, if not more so. These spirits range from relatively unknown mothers and fathers to Princess Diana and Harry Houdini.

A quick but essential clarification: It is important to note that I use the term *alleged* in regard to Spirit and that it is always implied whether stated or not. When I speak of Spirit, I am really saying alleged Spirit. This also applies to spirit guides, angels, or even the Sacred; each is implicitly preceded by the silent and invisible *alleged*. When I mention the continued presence of Professor Einstein or Princess Diana, the silent and invisible *alleged* is there as well.

The second category of evidence involves the possibility of spirit-assisted healing. The examples range from carefully examined case studies where Spirit appears to have played a role in healings (Type I), to proof-of-concept laboratory investigations (Type II) and experiments (Type III) demonstrating how the spirit-assisted healing can be scientifically investigated in the laboratory.

The third category of evidence is by far the most challenging and controversial. It involves the possible existence of a greater spiritual reality including angels and spirit guides. I include personal examples of addressing the Spirit guides hypothesis with an open yet critical mind, as well as the use of supersensitive biophysical instruments to potentially detect the spirits' presence.

Let's begin by raising the question, if Spirit exists, how can it prove its existence to us? As you will see in the next chapter, a formal criterion has been developed to pose such questions about thinking computers, for instance, none of which have passed the tests at this point. My proof of concept is that if spirits are proven to be as mindful and willful as we are, then they at least pass the test of the survival of personality . . . and possibly more.

PART II

The Promise of Spirit's Willful Intent

3

DOES EINSTEIN STILL HAVE A MIND, AND CAN HE PROVE IT?

Do you believe in immortality?
"No, and one life is enough for me."

—Albert Einstein

This may be hard to believe, but a number of research mediums, whom I use in afterlife laboratory experiments, regularly tell me that Einstein is alive and well on the other side, that he has important messages for humanity, and sometimes the receivers include me. If these mediums are correct, then the Einstein quote that introduces this chapter is no longer valid.

Although I listened quietly to these reports, until now I have done nothing about them. I have not written about the numerous, independently replicated claims of these mediums nor their emerging complementary evidence concerning Einstein's existence in the spirit world. Moreover, I have yet to invite Einstein to participate formally in our laboratory research.

My discussion of Einstein and his continued consciousness points toward a real scientific dilemma that needs to be approached thoughtfully and thoroughly. It provides us with the opportunity to explore what we consider evidence of Spirit's existence and why we need an expanded and intention-focused criterion. This criterion is important in not only establishing Spirit's existence but also helping establish whether our internal voice is a psychological projection or if it has a character of its own and may be the whisperings of Spirit guiding us.

The Meaning and Use of *Prove* and *Proof of Concept*

A point of clarification and definition: It is important to understand that scientists do not typically use the words *prove* and *proof*, and this includes me. Scientists prefer statements like "determine the probability of a given explanation accounting for the available data." The word *prove* is usually reserved for mathematics: "to verify the correctness or validity of by mathematical demonstration or arithmetical proof" (*Random House Unabridged Dictionary*).

However, I am intentionally using *prove* in this book in its more general, everyday dictionary meaning of the term: "to subject to test, experiment, comparison, analysis, or the like, to determine quality, amount, acceptability, characteristics, etc." (*Random House Unabridged Dictionary*). Also, as noted earlier, there is the proof-of-concept stage, an experiment, pretest, or even a prototype model, which demonstrates in principle the feasibility of a scientific approach or a product's viability to verify that the concept or theory is deserving of further or more formal experimentation or product development.

The proof of concept is usually considered a milestone on the way to a fully functioning prototype. I consider all of the preliminary and exploratory investigations and self-science observations included in this book as examples of proof of concept and proof in principle, because they are demonstrations of the potential use of the scientific method to address the questions, "Is Spirit real?" and "Can Spirit play a role in our personal and collective lives?" And by example, "Is Einstein still here?" and "Can Einstein play some role in our personal and collective lives?"

People often ask me, "Are you trying to prove X, where X is life after death or the role of Spirit in healing?" What I say is: "Absolutely not. What I am attempting to do is to give X the opportunity to prove itself." If X exists—in this case, the larger and more general idea of Spirit and a greater spiritual reality—we are using the scientific method to ideally optimize the possibility of Spirit to establish its own existence. In other words, the proof of the pudding is in the tasting, a precept that businesspeople as well as scientists, and mothers, use every day. So, if Einstein exists, what we could do is give Einstein the opportunity to prove it in the laboratory—that is, if we are brave enough.

But before we consider this new approach, it will help us to conduct a related Einstein *gedanken* experiment, or thought experiment—one carried out by proposing a mental hypothesis only. Einstein loved gedanken experiments; his most famous was imagining as a boy that he was riding alongside a light beam traveling at the speed of light. Our gedanken experiment requires that we imagine that Einstein is indeed still here, and we try to put ourselves in his spiritual shoes.

Putting Yourself in Einstein's Shoes (or Consciousness)

Let's imagine that Einstein's granddaughter goes to a medium. Let's assume for the moment that the medium is not a fraud—engaged in fake mediumship (doing what is called "cold reading," or using a mental magician's set of techniques). Also, the granddaughter is being a good sitter—the person for whom the reading is being given—meaning she is not giving the medium cues or information about her grandfather. In fact, let's imagine that she has not told the medium that she is related to Einstein; in other words, the medium is blind to this fact.

Einstein is aware—from his position on the other side—that just because the medium provides verifiable evidence of his past history, this by itself is insufficient to prove that his consciousness has survived bodily death to any objective observer. He and any objective observer understand that it is theoretically possible that the medium might be psychic and telepathically reading his granddaughter's mind. If the medium is reading her mind, the accurate information, supposedly coming from him, would not in itself be proof that it was from Einstein on the other side.

Even if the medium is not reading his granddaughter's mind, she might be obtaining the information from other sources, including what some scientists speculate might be the so-called zero point field (mentioned earlier). Just as information encoded in starlight continues to exist in the vacuum of space long after the star has died, it is theoretically possible that the medium is retrieving the information and energy Einstein left in the vacuum of space that was created while he was alive.

Let's further imagine that his granddaughter is quite skeptical—even worse, that she is a professional scientist like him. Einstein is well aware of the theoretical possibility of mind reading as well as what is termed *super-psi*—such as reading dead information in the quantum vacuum of space. How would his skeptical loved one know that Einstein, in spirit, was actually in communication with the medium? How would Einstein demonstrate to his skeptical loved one that he was actually conscious, alive, and evolving on the other side?

What Does It Mean to Have a Mind, a Living Consciousness?

To understand Einstein's dilemma, my deceased mother's dilemma, or any other spirit's, let's imagine that you are the deceased. How would you, on the other side, go about proving that you were not only still conscious but alive as well?

To address these fundamental and far-reaching questions, it is helpful for us to step back and consider what it means to be both conscious and alive in the first place, and examine how we would go about proving that we were conscious and alive in the physical world.

Let's consider consciousness first. How would I prove to you that I am conscious as I am writing this book?

The truth is that the only consciousness we know for sure is our own. I experience my thoughts and feelings; you experience yours. Though you may empathize with my thoughts and feelings, you cannot experience them directly. For example, I may tell you that I am seeing a beautiful orange sunset, and it may look orange to you too. However, you can't be sure that what you see as orange is the same experience of orange that I have, nor can I be sure that mine is the same as yours. Just because I tell you that I am conscious is not proof that I am conscious.

As a case in point, I can program a computer to engage in complex dialogues that sound very much like me.

You could ask the computer: What is your name?

I could have the computer programmed to detect the vocal pattern, "What is your name?" and it could respond by saying, "My name is Gary."

You could ask: Where do you live?

Likewise the computer program could detect this pattern, and respond with, "I live in Tucson, Arizona."

You might then ask a trick question: Are you conscious?

I could have the computer programmed to detect the pattern "Are you conscious?" and say, "Obviously. Are you?"

I could program tens of thousands of possible questions you might ask and connect them with multiple kinds of answers designed by me, and thereby make the computer seem quite conscious, creative, and alive. In fact, if we were conducting this communication experiment over the telephone, you might think that you were actually speaking with me!

The bottom line is that I can't prove to you that I am conscious, especially since contemporary computers could play me quite well. The take-home message is that the best you can do is to infer that I am conscious from my behaviors. And I must do the same with you: infer that you are conscious from your behaviors.

This fundamental challenge (some would say problem) is compounded after we die. If I can't prove to you, definitively, that I am conscious when I am physically alive, how can I convince you that I am still conscious after I die? The crux is that the medium must infer that the deceased spirit is conscious just as you infer that I, in the physical, am conscious.

It is essential to understand that experimental physicists are used to inferring processes they can't measure directly. They believe in the existence of a gravitational field because of the behavior of objects moving in space or the behavior of numbers changing on a computer screen. Gravity cannot be seen, heard, or detected with our primary senses. It is inferred indirectly by its effects on matter or light, and the same goes for consciousness. I know I am conscious, and I infer that you are conscious. You know that you are conscious, and you infer that I am conscious. Proving that we are conscious is difficult enough. Proving that we are a living consciousness is even more difficult.

Think about this: Obviously there is no difficulty in proving to you that I am biophysically alive. You can measure my brain waves, my heartbeat, my oxygen consumption, and so forth, and conclude that I am currently, physically alive. However, what would you measure to prove that my consciousness is alive? Moreover, how would I prove to you that I am mentally alive?

On the face of it, this may seem obvious too. You can ask me questions and I can answer them. But does this process prove that my consciousness is alive? The computer-programming example presented above—sometimes called the Turing test after the computer scientist who first described it—reminds us that the question-answer paradigm does not prove the existence of consciousness, let alone aliveness. Although creative and clever responsiveness is consistent with aliveness, it is not proof of it.

When a medium says, "I see Einstein, and he is jumping up and down," does this necessarily mean that Einstein is actually jumping up and down on the other side? Or could the medium be seeing a historical record of Einstein jumping up and down, if not actually *imagining* Einstein jumping up and down?

Just because the medium says that she sees Einstein doing or saying specific things does not necessarily mean that Einstein is actually there and alive, even if the information is highly accurate and independently verifiable by other living relatives or scientists. The issue here is not one of accuracy; it is of survival of consciousness and aliveness.

So how does Einstein on the other side prove that he is still a living consciousness? How does his granddaughter, who in our thought experiment happens to be both a skeptic and a scientist, come to the conclusion that the great scientist—his very essence—is still here? And it doesn't matter whether it is Einstein the deceased scientist or Shirley my deceased mother—they both have the same problem. And now conceptually you do, too, if you feel you're having genuine contact with a deceased loved one or spirit guide.

Showing Scientists that the Deceased Have Minds of Their Own

This has been the ultimate question for scientists and afterlife researchers interested in documenting whether consciousness survives physical death. It is relatively easy for researchers to establish definitively that conventional explanations of mediumship such as fraud, cold reading, rater bias, even experimenter bias, cannot account for all of the findings. It is more difficult to establish that the medium isn't reading the conscious mind of the sitter.

As I discussed in my previous books, there is substantial evidence that mediums typically fail to relay the precise information that the sitter is consciously thinking about, or provide information that the sitter has forgotten and recalls later, or provide information that the sitter never knew in the first place. Other professional scientists, such as Dr. David Fontana, as well as lay scientists, such as Susy Smith, have discussed this in their books as well. This still leaves open the possibility that the mediums were doing something that is superpsychic, such as reading the distant mind of a relative or friend, or tapping into information left in the quantum vacuum of space.

Meanwhile, remember that we are imagining in our thought experiment that Einstein happens to actually be there on the other side, and that he is trying as hard as he can to convince his loved one—and you and me, for that matter—that he is not only conscious but very much alive as well. How does he do this?

The answer is that he asserts himself, and he does so creatively and convincingly.

Spirits Assert Themselves in Evermore Ingenious Ways

The truth is, though I was familiar with the research spanning more than a hundred years addressing the survival of consciousness hypothesis, I did not anticipate that new convincing evidence would begin to show up unexpectedly, and repeatedly, in both my formal laboratory experiments as well as my informal self-science investigations, through no conscious effort or control of my own.

From the very beginning of my research program, I witnessed unanticipated and unplanned instances that were strongly suggestive of the survival of Spirit. This evidence was accentuated, sometimes extraordinarily so, following the deaths of distinguished persons who were scientists in their own right—especially Susy Smith—and who were very familiar with afterlife research. You will meet some of these people and the lessons they taught me in the next four chapters.

As I witnessed these accumulating instances of apparent survival, one after the next, it became clear to me that the way for deceased people—such as Einstein and my mother—to prove that they were alive is simply for them to assert themselves.

They could take charge of the situation and show that they had qualities of conscious intention—what we scientifically refer to as discarnate intention.

Basically, the deceased Susy Smith could establish that she has a mind of her own, and she could use it creatively, sometimes playfully, and in certain instances, definitively. She could show that she has at least as much freedom as you or I do, that she could make choices and march to her own drummer.

The deceased could even establish that she is not a slave to our experiments, but rather she is as much in charge of the outcome as, if not more so than, the scientists conducting them.

In other words, the deceased could demonstrate that the medium is not getting the information; rather, the information is being given to the medium. As Dr. Julie Beischel and I expressed in our 2007 paper published in *EXPLORE: The Journal of Science and Healing*, mediums do not retrieve the information; they receive it.

Curiously, my emerging research reveals that people who regularly took charge in life continue to take charge in the afterlife. People who asserted themselves on this side continue to assert themselves on the other side. Those who were creative and cunning on the earth appear to behave creatively and cunningly after they died.

Remember, this criterion also applies to your internal voice: is it your psychological complex talking to you, or something independent of your own history that could be "other-directed," as we say in afterlife research?

Using scientific language, the information received by mediums revealed properties that strongly implied the existence of: intention, assertion, decision making, self-control, disagreement, stubbornness, and so forth—characteristics that people showed in life that continued after they died.

Let's consider one last key question before we explore the proof-of-concept evidence.

How can I prove to you that I have a mind of my own—and by extension, how you can prove to me, or your loved ones, that you have a mind of your own—regardless of whether you are in the physical or in the post-physical, on the other side?

If I choose:

1. I can interrupt you when you are speaking.

2. I can surprise you with an unexpected visit.

3. You can ask me one question, but I can decide to give you an answer to the previous question.

4. You can make a statement, and I can disagree with you.

5. You can ask me to stop, and I can decide to ignore your request.

6. You can ask for a specific piece of information, and I can say, "No, I won't tell you."

7. I can even choose to lie to you.

If survival of consciousness is real, then theoretically, I can do these things whether I am in the physical or not. If you give me the opportunity to assert myself, I can show you that I am my own person.

You can do this, too.

So can Einstein.

Emerging Unplanned Evidence for Discarnate Intention

The next four chapters detail a series of surprising instances, mostly unplanned and unanticipated by myself, that have occurred in the context of our afterlife experiments research. I confess my confusion, surprise, delight, disbelief, frustration, and wonder as each account unfolds. These unplanned instances include little-known deceased people like my mother and mother-in-law and well-known deceased individuals such as Princess Diana. As one of the deceased scientists taught me after he had died, "Survival is in the details."

And as you are about to discover, the deceased are sometimes more crafty and compelling than we scientists could ever be—and a lot more mischievous.

If the deceased really have survived, then they obviously know more about the afterlife than we do.

The question is, are we ready and willing to listen? Are we willing, with their guidance and collaboration, to design and conduct the kinds of future experiments that can definitively establish, beyond any reasonable doubt, that they are still here?

As challenging as it is scientifically to establish that they are still here, it is even more daunting to establish that it is their fervent intention to help us, individually and collectively. Yet this is precisely what science must do: find a way to document that Spirit intention not only exists but that it is focused on our welfare.

This makes the scientific case relevant to my exploration of Spirit and yours. If we're going to accept advice from Spirit or follow through with an alternative practitioner, we better be sure that this is real and we're not just fooling ourselves or projecting our hopes and wishes into the vacuum of space. *The Sacred Promise* is about this great possibility; it provides the essential proof-of-concept evidence for taking this work forward in a serious and responsible manner.

Remember, if the research mediums are correct, Einstein and other great historical figures are semipatiently waiting for all of us, including conservative scientists, to finally take them and their intentions seriously. And as you will see, the perceptions of these responsible and devoted research mediums—as strange as they may sometimes seem—deserve our serious consideration.

4

CAN NECESSITY SUMMON
AN INVENTIVE MOTHER?

Mother is the necessity of invention.

—Anonymous

It is sometimes said that necessity is the mother of invention. Growing up, my mother taught me the reverse: mothers are the necessity for invention by their children. Over the years my mother helped me understand the big picture in many surprising and memorable life lessons, often using her creativity and playfulness but always doing things her way.

Mothers-in-law can be teachers, too, even after they die. In this chapter I am going to present evidence from two investigations: one conducted in the research laboratory involving my mother, Shirley, and the other in the laboratory of personal life involving my mother-in-law, Marcia Eklund, and her daughter, my wife, Rhonda. They illustrate in clever and convincing ways that the deceased can assert themselves and prove that they have minds of their own as they help us.

You may be wondering, in what sense are such observations scientific? Moreover, you may be pondering if they reflect a subjective bias on the part of the observers. These are valid and fundamental concerns that deserve to be addressed at the outset.

First, the observations reported here are scientific in the sense that they were recorded carefully and in some instances were witnessed by two or more individuals. These events really happened. There is no fraud or misrepresentation here.

Second, the observations were analyzed from different perspectives, including careful questioning of the context of the events as well as their possible interpretations. I am a stickler for considering alternative explanations of observations and giving all possible interpretations a fair hearing.

Third, extensive research in both quantum physics and parapsychology points to the fact that the openness and intentions of the observers—be they a professional scientist or a science-minded layperson—can affect what is being observed and measured. Openness to discovery as well as appreciation of what is being shown may create an optimal state for the emergence of certain genuine phenomena, while multiple observers rule out self-deception. This also applies to the self-science investigations I conduct here; a friendly show-me attitude, like mine, seems to summon a positive response from skeptics who may try to replicate the experiments, while anything less may send them packing.

And finally, in the early stages of scientific research in a new area, establishing the viability of innovative concepts as well as methods to investigate them are essential to justify and design future systematic research in the area. It is in this context that I share these important, personal proof-of-concept observations.

Lessons from Shirley's Big Picture

Most of our mothers are not well-known, save to their children and their small circle of family and friends. However, our mothers typically have a foundational and lasting impact on our lives, and as I have come to discover, they can continue to attempt to mother us even after they have passed.

Although my mother was an unknown classical pianist turned elementary school teacher—her only brief public notoriety was serving as president of the parent-teacher association in our Long Island school district—she was bigger than life to those who knew her. She was known to be a good person who had a huge heart, overflowing with caring, courage, and conviction. Family and close friends experienced firsthand that she was a force to be reckoned with; my mother not only garnered our love and respect, but sometimes our wariness as well. We learned that it was not wise to cross her.

As mentioned in the previous chapter, my afterlife research (and that of other investigators) indicates that people who are impactful in this life generally continue to appear to be as big in the next. It is curious and probably fitting that the first clear evidence for discarnate intention was taught to me by my deceased mother.

Parts of the incident I describe below were revealed in *The Afterlife Experiments*. However, it is now time to present the complete (and controversial) story and the deeper lesson of discarnate intention that my mother tried to teach me. I say "tried" to teach me because it has taken almost a decade of creative replications of her intent by other deceased parents and children—witnessed by me and independently by my colleagues and other scientists—for me to finally accept what my mother was apparently showing me. Of course, this also implies that my mother's consciousness and intentions have survived.

I would like you again to participate in a thought experiment, in this instance trying to imagine that you are my deceased mother.

How would you convince me, the agnostic/questioning scientist, that your consciousness and personality, for better or worse (only kidding, Mom), were still alive and well? Especially when I had specifically asked her—in my head (allowing for the possibility that she might be still here)—not to interfere in a specific controlled experiment I was conducting at that time!

It was the summer of 1999, and my colleagues and I were about to conduct a carefully controlled investigation with three research mediums and five research sitters. This was the first experiment we had designed where the sitter, the person receiving the reading, was completely silent—she was never allowed to speak. The research was funded through a gracious gift from Canyon Ranch—the gift included travel and lodging expenses for the three mediums and the use of space at the ranch to conduct the experiment.

Each medium worked with a particular investigator, or conductor. I realize now that I had the good fortune to serve as the experimenter working with the then-relatively unknown medium John Edward. During the experimental sessions or readings, the mediums sat in separate rooms facing video cameras, with their backs to floor-to-ceiling screens that separated them from the sitters and experimenters. The mediums were allowed to talk only with their respective experimenters, and never allowed to speak with the sitters.

The research sitters, all female, had some trepidation with an extensive postexperimental analysis. After they had individually experienced separate readings with each of the three mediums and we had had the tapes transcribed and prepared for item-by-item scoring of the responses, they would then have the daunting task of carefully rating all of the readings. Not only would they be required to rate each and every item from their personal three readings using a 7-point −3 to +3 accuracy scoring, they would also have to rate each and every item from the other twelve readings of four sitters, as if the information applied to them. Our purpose was to determine how unique the information from a given medium's reading was to a specific sitter as compared with control readings, i.e., readings of other sitters of the same sex.

Each experimental reading consisted of two parts. In the first part, the medium was asked to report whatever they could get about the sitter's deceased loved ones. The medium was not given the name of the sitter nor told the names of the deceased loved ones. The medium was not allowed to ask any questions of the sitter, and the sitter was not allowed to speak.

In the second part, the medium was allowed to ask simple questions of the sitter that would allow yes or no answers. However, the sitter was still not allowed to speak. What she could do was to shake her head yes or no, and the experimenter, who was blind to the sitter's histories of their deceased loved ones, would say yes or no. The only voice that the medium heard during the total of five readings (one per sitter) was the voice of the experimenter. Hence, the only voice that John heard during the running of the experiment was mine.

It occurred to me approximately a week before the sessions were to be conducted that a clever addition to the experimental design would be to add one more reading, only this time there would be no female sitter. Instead, the experimenter would enter the room without a sitter, and would serve as a secret sitter himself or herself. (It appears in retrospect that I wanted to get a reading of my own, or was being prompted, as you will see.)

I say secret because the mediums would not be told that one of the readings would be without a research sitter. The mediums would be kept blind to this addition. This would increase the total number of readings from five to six, and therefore the total number of readings scored by each of the sitters from fifteen to eighteen. More-over, each of the three experimenters would be required to score all eighteen readings to determine if unique information would be obtained when the experimenters were the secret sitters.

Unfortunately, when the mediums arrived in Tucson and learned that we were planning to hold six readings during the experimental day rather than the initially agreed upon five, they balked. They complained, rightfully so, that they would be well fatigued after five readings. They reminded me of the psychological and physiological stress of having to do one intense reading after the next, compounded by the fact that they were not being allowed to see or speak with the sitters.

I realized that I could not, in good consciousness, deny one of the five previously selected research sitters the opportunity to participate. They were each so looking forward to having these special research readings with such a distinguished group of research mediums. After conferring with my research team, I decided to forgo the extra sixth reading where the experimenters were to serve as secret sitters. This meant that experimenters' deceased loved ones would not have the opportunity to participate.

Having completed two previous experiments with John, I had witnessed profound evidence not only that he was genuine, but that the information he received supported the conclusion that conscious-ness survived physical death. So I disinvited my mother, father, and grandparents from participating in the experiment. However, I prom-ised them that in a future experiment, I would serve as a secret sitter and that they would have the opportunity to serve as an officially departed loved one in our research. Since I am not a medium, I had no idea whether my deceased family members heard me and that my mother apparently had her own plans. She was going to show up at the end of the day and, in the process, make sure I realized that she was still the boss—or at least, her own boss, and that she had some-thing important to show me now.

Shirley Barges into the Experiment

As we reached the fifth and final reading of the day, John and I both were exhausted and looking forward to wrapping up. Nonetheless, John went ahead with the last reading, picking up images and names of people who the sitter recognized and confirmed during the yes-no portion of the reading. But it wasn't long before John acknowledged a difference in what he was experiencing. It is important to pay close attention to the precise wording John used:

> John Edwards (JE): This is not flowing like in my normal conversational style; it's being given to me in, like, big blurbs . . . kind of like what I wrote down on the paper before everybody came in.
>
> They're telling me that the female S-sounding name is here, acknowledging her boys. One must be in the medical field 'cause he's a doctor. That she has her husband there. She talks about the sign of Gemini, which either means somebody's a Gemini or is a twin. But that's not for the sitter. Gary, it might be for you.

I was both stunned and secretly pleased. My mother, whose name, Shirley, started with an S, as did her middle name, Sarah, did have two sons, one of them was a doctor—me, though not a physician but a PhD, and I am a Gemini. These four facts defined a specific pattern that fit me perfectly. But was the information really for me? John continued getting more information for the same person.

> JE: And somebody wants to be called "the milkman." And that's weird because he's not trying to show me that he delivered milk. He's the milkman . . . I have two moms here. They're not related at all. Gary, for you . . .

Two different moms, not related at all? Startled and intrigued, I encouraged John to continue with what he was receiving, regardless of whether it was for the sitter or me. More factual information was revealed. More of it occurred during the yes-no period, and since John identified this exchange being for me, the answers that I gave were my own, not for the sitter.

JE: . . . was your gallbladder removed?
Gary Schwartz (GS): Not it specifically.

JE: I'm sorry? Was there stomach surgery like gallbladder, appendix removed?
GS: Yes.

JE: And your mom has passed?
GS: Yes.

JE: Is she the S?
GS: Yes.

JE: Do you have a brother?
GS: Uh-huh (yes).

JE: I am not sitting with your brother, correct?
GS: Excuse me?

JE: This is not your brother?
GS: No.

JE: Okay. This is for you. The milkman is for you.
GS: Hmm.

JE: Your dad also passed.

GS: Yes.

JE: Is there a Morris in that family?
GS: Yes.

JE: What I see is it being like an uncle or a grandfather.
GS: Uh-huh (yes).

After this dialogue, John requested permission to ask my mother to be quiet so he could receive information for the sitter. But he continued to have trouble. And then he identified the source of the problem; as he had suggested earlier, two mothers were present.

JE: One for the person, one for you. Your mother is louder, Gary.

John's interpretation here was consistent with my mother's personality. She had often been the dominant person in a conversation, to put it mildly. John was then able to focus on the actual sitter of the session, and although he received significant and accurate information for her, he continued to report information that was accurate for me as well.

He confirmed that the milkman he had mentioned earlier was for me, and most interestingly, he claimed that my Uncle Morris was known by two other versions of his name which sounded like Maurice or Merle. Two other versions of his name?

After the experiment, I called my brother who in turn called our cousin, Uncle Morris's son. The cousin confirmed a version of what John had said: his father was sometimes called Moshia and sometimes Moe. Not the exact two names that John had given, and it is debatable whether his versions of the names were close enough or not, but about the man being called by two names other than his own that both began with the letter M, this was right on target, and I did not know this at the time.

If I applied our standard −3 to +3 scoring procedures to the information John said specifically came from my family for me, my rating would be at least 80 percent +3 accurate. The unique combination of information—mother's S name, one son in medical field, a Gemini, appendix out, a brother, a deceased father, Uncle Morris with two or more M names (this had no correlation to any of the sitters or the other experimenters)—applied only to me. The probability of this pattern of information occurring by chance is much less than one in a million.

Most of the information about my family was interesting and even accurate, but what about the reference to the "milkman"? When I checked with all the sitters and experimenters to see if any of them had a relationship or connection to a milkman, none did. But I hadn't expected any of them to report a connection; I was pretty sure from the outset that the information was probably for me. Indeed, it brought back a fond childhood memory, something I had not thought about in decades.

As a youngster, I had developed great enthusiasm for collecting glass milk bottles, which I used for storing the kinds of things little boys tended to collect in those days: favorite plastic cowboys and Indians, old pennies, assorted deceased Japanese beetles, and lightning bugs. I had a large collection of these bottles, which must have put me on a special blacklist with our friendly milkman, who was constantly asking for them back, to my continued refusal. The day milk bottles and the neighborhood milkmen were replaced by milk cartons from the grocery store was a jolting one for this young child, as well as for his sympathetic and caring mother.

Was John's mention of a milkman a specific reference to a childhood memory of mine? Obviously we can't know the answer for sure. But I was inclined to entertain that possibility. Would my mother have purposely brought this up in the context of apparently barging into this experiment? It was only much later that I realized another implication of this milkman reference. At that point in my career, I was beginning to conduct exploratory afterlife investigations but had not yet committed to writing about them. Knowing my tendency, was my mother trying to tell me not to hoard this information, as I had my assortment of toys? It would be dangerous for me to go public with this information—especially discoveries made in the laboratory of my private life—and she knew it. Was this her version of the divine boot?

Interestingly, in the hundreds of research readings I have been privileged to witness since then, no deceased members of my family have ever barged in. They have never taken over a reading, even for a brief period of time. And yet, the one time I secretly invited my mother to participate in an experiment, and then disinvited her, she appears to have shown up anyway. As you will see in the instance I am about to reveal, my mother sometimes had—and apparently she continues to have—a clever and irreverent sense of humor.

A Subsequent Secret Informal Reading Involving My Mother

After witnessing my mother seemingly drop into the Canyon Ranch experiment and monopolize the fifth reading for a while, I felt the need to serve as a sitter in a secret reading with another medium who was unaware of what had happened in the surprising Canyon Ranch experiment.

One night the medium contacted me claiming that he sensed my mother's presence and that she wanted to speak with me. I asked the medium if my mother wanted him to do a reading for me. The medium said yes.

Because this was not a formal investigation or experiment—it was a personal reading—the reading was not recorded; also, I did not take notes at the time. However, because the information was so specific, meaningful, and memorable, the experience feels like it could have happened yesterday. Also, shortly after the reading, I shared the information with a few people, including my brother.

I asked the medium if my mother could specifically show us how she died. The medium said that my mother was showing that she was seriously ill, and that she died from a long-standing illness. This was correct; my mother had suffered from diabetes and high blood pressure, and she eventually went into kidney failure.

I asked the medium if my mother could give him information about the funeral. The medium said that my mother was showing the service, and afterward, a funeral home where I picked up my mother's ashes. This was true; my mother was cremated.

He then said, and I paraphrase, "Your mother is showing me that her ashes were in something like a shoebox. The container was crude."

I was stunned by the medium's comment. At that time, I had never gone to a crematorium before to pick up someone's ashes. Since I did not request a fancy container, my mother's ashes were put in a plain coffee can! Only my brother and I knew that her ashes were provided this way. Though the medium did not say coffee can, it's being a crude container, something like a shoebox, is noteworthy.

The medium then said, "Your mother is showing you and your brother taking a ride to the water. And that it was very cold and windy." This was true. My mother died in the winter. She had requested that my brother and I spread her ashes in the Great South Bay area off the coast of Long Island. My brother and I drove to a beach area near the Captree Fishing Dock to spread her ashes. It was a bitter-cold, gray, and snowy day.

The medium said, "Your mother is showing me that you had a difficult time spreading her ashes. That they were blowing in the wind." Now I was shocked.

The bay had been partly frozen, and my brother and I had to walk out on the ice to get near the water. When I tried throwing my mother's ashes in the water, the wind not only blew them back in my direction, but blew them on my pants and jacket, and even in my face.

My brother and I felt not only exceedingly foolish doing this but also unnerved, yet we were attempting to honor my mother's last request—at least so we thought. Remember, my mother died in the early 1980s, and my brother and I had been taught to believe from childhood that it was "ashes to ashes, dust to dust, case closed."

And then the medium gave me one of the strangest and most comforting messages I ever heard a medium offer. He said, "Your mother is showing me that both she and your father were there." My father had died a few years before my mother. "She shows herself and your father smiling, and she is telling me that both she and your father thought what you and your brother were doing was quite humorous." The experience had actually been traumatic for me, being covered like that in my deceased mother's ashes, almost like being suffocated by her sometimes overwhelming presence. The medium's message helped heal that wound.

However, no matter how convincing the Canyon Ranch experiment or its observations were, the hallmark of science is replication. I would have to witness evidence of deceased people appearing to drop into our research again and again (and again), before I would finally give up my well-honed skepticism and conclude that my mother, my mother-in-law (described next), and by extension, all of our loved ones were still here.

What is curious is that as time went on and the research unfolded, the evidence for discarnate intention became ever more unexpected, seemingly unbelievable, and increasingly incontrovertible.

It was as if the more prepared I was to envision and make some sense of the emerging evidence for discarnate intention, the more elaborate and inventive the evidence for intention would become.

Was I becoming personally interested in the discarnate intention hypothesis? Yes.

For this reason, I was mindful to pay close attention to possible nonintention (and therefore nonsurvival) interpretations of the emerging findings as well.

Lessons from Marcia after She Died

The information I am about to share below comes from my wife's diary about her personal experiences of afterdeath communications following the passing of her mother. It is presented in greater detail in her forthcoming book *Love Eternal*.

I am sharing this one incident here because it reveals another remarkable way that spirits can assert themselves and prove not only that they can still be with us but also that they can apparently choose to help us at appropriate times in our lives—the essence of *The Sacred Promise*.

When I met Rhonda, Marcia had been deceased for more than five years, and Rhonda had produced a carefully typed report of a progression of extraordinary incidents that led her to the conclusion that her mother's spirit was still here. Rhonda was an only child, and she and Marcia had had an exceptionally close relationship. Marcia was a deeply religious woman who practiced spiritual healing later in her life. (This fact will become important in part III of this book, when we discuss emerging research on spirit-assisted healing.)

Here is the account, in Rhonda's words, from her report:

Monday, November 26, 2001
My days over the next week were filled with tending to affairs involving settling the small estate. I made an appointment and went to the company my father had retired from to inform

them of my mother's passing. She and I had met with them after he had become disabled. At that time there had been a ten-year certain annuity policy created which would now need to be referred to change the distribution.

They informed me that they could not find their paperwork and that they didn't remember any such agreement.

During this time, I was continually praying to God and talking to Mom. I returned home and remembered thinking that my nature would normally have been to riffle through every drawer and file in the house again, bound and determined to find our copies of those documents. [Note: Rhonda had previously searched the house and failed to find the policy]. Needing to be back at work in Seattle the next week, however, I didn't have the time to do that.

I stood in the archway of our house between the living room and dining room looking up and said, "God, I know you know where those papers are and you can reveal their location to me. And Mom, I know you know where they are, too, and why I need to find them, so if you can show me where they are, please do."

Immediately, the thought came to me, as words being spoken to me in my head: "Go into the closet in my office and look behind Dad's folded flag on the floor."

I didn't question or analyze the directive. It was easy enough to check.

I walked into the office with curious expectancy, opened the partially closed door, and felt my heart start beating faster as light from the room flooded the closet, revealing the folded military flag leaning against the wall with some loose papers sticking out from behind it!

I couldn't believe my eyes. Could it be?

I picked them up, my thoughts racing, not yet really having the time to fully consider the implications if they were the papers I was looking for.

Yes, they were the papers I had unsuccessfully turned the house upside down looking for.

Of course the Marcia account, as dramatic as it is, does not establish definitively that lost objects can be found, nor does it establish definitively that Spirit was actively involved in the process. But had Rhonda been unable to find this annuity policy, it would have had grave economic consequences for her and her emotional state of mind and need, as the situation has been in other intervention scenarios with Spirit. This is another consistent aspect of Spirit's apparent desire to help us.

Parapsychologists will be quick to point out—and appropriately so—that lost objects can sometimes be found using a technique termed "remote viewing." However, what is important to understand is that some people, some of the time, can find missing objects using remote viewing procedures tells us nothing about *how* the remote viewing works.

For all we know, it is possible that the remote viewing of missing objects works because Spirit sometimes assists the viewer in locating the missing object, whether the remote viewer is aware of the assistance or not. The fact is that we do not know the mechanisms of remote viewing.

Here's what's scientifically important, the big picture or take-home message: The combined examples from Shirley and Marcia in this chapter reveal the opportunity that we have to bring such possibilities—drop-ins and finding lost objects—into the laboratory and put them to experimental test. If we are not open to the idea that Shirley and Marcia are really here, we will never conduct the kinds of research that can verify whether or not the deceased are still here and can be of assistance in our lives.

THE SACRED PROMISE

You are about to discover an example of a sophisticated deceased person implementing a novel experimental paradigm, proving that one deceased person can choose to bring another to a medium—even when the medium is completely blind to the identity of the second deceased person and doesn't even know that the first deceased person is coming for a reading. And doing so all to prove one point: we're here and we want to help.

5

THE HELPFUL SPIRIT GO-BETWEEN, SUSY SMITH

Winter, spring, summer, or fall.
All you have to do is call.
And I'll be there, yes I will.
You've got a friend.

—James Taylor

When I think of the late Susy Smith, author of some thirty books about parapsychology and life after death, I think of James Taylor's song "You've Got a Friend." As it turned out, she was a friend not only to me but to all those seeking verification of life after death and of Spirit's willingness to help and guide us.

I have previously shared how I met Susy, and I have described in detail some of the research we've done together, before and after she died, in my previous books. In brief, when I first met Susy, she was eighty-five years old and preparing to die. She had even written her own obituary. Susy quickly became my first (and only) adopted grandmother; she used to playfully call me her illegitimate grandson. And as Susy was fond of saying, she couldn't wait to die so she could prove to the world—including me—that she was still here.

However, I could never have foreseen, or even imagined, the creative and extraordinary proofs that Susy would subsequently provide from the other side about life after death and the active role that spirits can continue to play in our lives. As part of my personal quest to discover if Susy was still here and would continue to play a meaningful role in my life, she ended up revealing a new research protocol for exploring the apparent reality of Spirit and its possible benefits for all of us.

I continue to find it remarkable how what I do in my personal life—particularly related to Spirit—ends up seemingly informing and advancing my professional concerns. It could be because the mind or ego separates professional from personal as it does alive from deceased, while Spirit insists on one unified whole or continuum.

Curiously, my biological mother, Shirley, and my adopted grandmother, Susy, happened to share certain key qualities in common: a passion for life, an inherent goodness, a deep commitment to their respective causes, and a bigger-than-life personality that was independently and repeatedly detected by multiple mediums.

Like Shirley, Susy was tough. If Susy needed to assert herself, she would. And she was not about to let me misinterpret her beliefs, whether they were valid or not. Since this was her persona while physically alive, there was good reason to anticipate that, if her consciousness did survive death, she would continue to assert herself in this way.

Furthermore, if anyone was going to insist, after they died, that I not misinterpret genuine communication from the other side as being explained by telepathy, or that mediums didn't always read dead information from the vacuum of space, it would be Susy.

This would especially be the case when the deceased in question was Susy and I happened to be the sitter! If Susy could find a way to prevent me from misinterpreting what was really occurring, she would do it. Her pragmatic approach is a model for everyone seeking the truth about afterlife contact and how not to fool one's self.

Well aware that I was an orthodox agnostic, Susy knew it would take a lot of evidence—both in and out of the laboratory—for me to conclude that her consciousness survived and there is continued presence of Spirit in our lives. So Susy provided the evidence, in spades.

Surprise, Surprise, I'm Here

As I describe in some detail in *The Truth about* Medium, within twenty-four hours of Susy's death—she was eighty-nine—on February 11, 2001, I began conducting blind readings to see if mediums could obtain evidence of Susy's continued existence. These were not university investigations or experiments; they were private investigations conducted in my personal life, or self-science, as I call them.

Though these readings, on the whole, were remarkable for their accuracy and specificity, they did not in and of themselves prove that Susy's living consciousness continued after her death.

Moreover, if the tables had been turned—if I had died and Susy were still here—she would not be convinced by similar information from mediums concerning my survival either!

However, within a month or so of Susy's passing, I received a surprise email from a then-unknown purported medium. To honor and preserve her anonymity, I will call her Joan and say she lived in the Pacific Northwest. Joan claimed that she had been psychic since childhood, and because she was happily married with a young son, she strived to live a normal, nonpsychic life most of the time. However, deceased people still showed up unannounced in her house every now and again—sometimes she would know them and other times they would be strangers.

She then reported that a deceased older woman named Susy was hanging around her house, and that Susy had important messages for me. As I recall, Joan knew about Susy Smith, since I had described some of my research with Susy in my first book, *The Living Energy Universe*. Joan further claimed that she did not know if the information was valid, but she would like to learn what, if any of it, was correct.

I wondered: Was it really possible that the deceased Susy Smith had dropped in unannounced in a house in the Pacific Northwest and was actually badgering a gifted psychic housewife to make contact with me? I also wondered if the woman claiming to be psychic was crazy, or was she pulling my leg? At this point I didn't know anything about her.

Did Joan have ulterior motives? Was she seeking fame, money, or something else? I even considered the possibility that Joan might be a spy for the infamous Amazing Randi, who has a reputation for trying to expose scientists and others investigating the paranormal. Having witnessed his antics and deceptions, I would have been imprudent to simply dismiss such a possibility out of hand.

So, you may ask, why pursue it at all? If Susy was in fact alive and well and wanted to communicate with me, having someone with her background cooperating in my research from the other side would be invaluable.

I wrote Joan, thanking her for contacting me, and suggested that she share whatever information the purported female spirit claiming to be Susy Smith wanted to communicate. I told Joan that the only way we could determine its accuracy was for me to carefully examine and score it.

Joan responded positively to my suggestion and sent detailed information she supposedly received from Susy. As I read the email, I determined that more than 80 percent of it was not only factually accurate but it sounded like Susy as well. As I studied the information carefully, I realized that the specific content related directly to me could be separated into two categories:

The first category I called "watching-over" information. This was about me in my present life, of the sort like, "Susy shows me that you recently experienced X." Of course this was important to our overall exploration of whether Spirit can and will help and guide us.

The second type was what I termed "predictive" information. This pertained to me in the future, of the sort like, "Susy is showing me that in a few days Y will happen to you." This is somewhat relevant to the above-mentioned consideration and proved to be critical in one instance.

Of course, I still had no idea at this point whether Joan was legitimate. I thought that if Susy was actually watching over me and even providing me with predictive information, this could be tested experimentally by spontaneous as well as intentionally planned behavior. In a follow-up email, I proposed to Joan that we conduct an informal investigation with me acting as a private person, who happened to approach his personal life as a scientist, and he wanted to know whether this was Susy or not.

I suggested that five mornings a week, Monday through Friday, Joan would agree to contact Susy and ask her two questions:

1. What had Susy witnessed as happening to me in the previous twenty-four hours—the watching-over information.

2. What did Susy foresee happening to me in the upcoming twenty-four hours—the predictive information.

Joan would then email the information she believed that she had received from Susy. Later in the day, I would score the information item by item. It would be relatively easy to remember much of what I had done in the previous twenty-four hours. Since most of the information turned out to be in the watching-over category, we focused our attention on it. I would email Joan my scoring each evening so she would know, on a daily basis, how she was doing.

We agreed that no readings would take place on the weekends to allow Joan to take a "Susy break." Meanwhile, Susy presumably had to watch over me on Sundays, so she would be prepared to be interviewed by Joan the following Monday morning. This procedural detail will turn out to be important later.

Though Joan admitted she was nervous, she was eager to give it a try, and we decided to begin our informal personal experiment the following Monday.

Watching a Baseball Movie while Eating Chinese Food with Chopsticks in Bed

As you might imagine, I too was also nervous, but for different reasons. If Susy were really watching over me, and if Joan were a genuine psychic, then I might soon have evidence consistent with the hypothesis that the deceased, or at least Susy, could choose to be with us on a fairly regular basis, whether we are aware of it or not.

I knew that it would not be evidential if Joan simply reported that Gary was brushing his teeth, taking a shower, or driving to work, etc. This kind of prototypic detail would fit most people, including me, and hence be scientifically useless—even if Susy happened to be witnessing me doing these daily activities.

But if I purposely did novel things and experienced rare events and Joan reported these activities supposedly via Susy, this kind of unique reporting could be meaningful. To give Joan and Susy the best opportunity to succeed, I decided to begin the watching-over experiment that Sunday night by watching a novel movie in a novel way.

I watched a VHS tape of the movie *Field of Dreams*, released in 1989, which I had not seen in a few years. This movie happens to be related to the question of life after death; it even includes scenes in which deceased baseball players play on a baseball field built on a farm in the Midwest.

It is important to note that I rarely watch baseball. Though I appreciate the game aesthetically, the sport does not capture my interest or hold my attention. This fact will become important shortly.

I further decided to order in Chinese, something I had not done for two or three years prior to this watching-over experiment—I usually eat out at Chinese restaurants. In those days, I typically ordered in Italian food.

And finally, I watched the movie and ate my Chinese dinner in bed—three things I had never done together. As I learned that night, eating Chinese food with chopsticks while reclining in bed is neither easy nor advisable!

The following morning, with some trepidation, I turned on my computer and discovered an email from Joan: "Susy is showing me something about baseball; you were watching baseball last night."

Joan continued, "Susy is showing me that you are eating some kind of foreign food." She did not use the word *Chinese*. Although I had been eating foreign food the previous night, I could have been eating Italian instead, and the word *foreign* would have applied as well. Since I ate foreign foods at least two nights a week, I did not give this specific information much credence, although eating Chinese food at home was certainly a novelty. And eating with chopsticks was more foreign than using a fork.

However, what came next strongly captured my attention. Joan wrote, and I paraphrase, "I am trying to see where you are eating dinner. Susy is not showing you sitting at the kitchen table or the dining room table. I don't understand this, but she shows me that you are eating while reclining. Does this make any sense?"

The key word here is *reclining*, for that was exactly what I had been doing. I had been reclining in bed, eating a foreign food with foreign

71

utensils while watching a baseball movie. I could not recall the last time I had eaten dinner in bed watching a movie, and certainly not Chinese with chopsticks.

When I had scored all the information, the watching-over experiment for the first day turned out to be approximately 80 percent accurate.

Encouraged by the early success, we continued our personal exploration five days a week for more than two months. The information averaged around 80 percent accuracy. I carefully monitored my activities and found no evidence that I was being watched over by conventional means. I would not tell Joan when I would be traveling or where. If Susy were truly watching over me, I reasoned that she would be able to follow me wherever I went.

Sometimes the information provided by Joan was truly uncanny, especially when predictive information would show up in the form of warnings. One morning Joan wrote that Susy claimed that I needed to check the tires on my car. I did not act upon her purported request. Later that afternoon, I discovered in my laboratory's parking lot that my car had a flat tire. I could not remember the last time I had had a flat. I decided to have all four tires replaced.

The skeptic would be correct in raising the possibility that Joan might have been cheating. It is conceivable that she secretly arranged for someone to watch my activities, or even let the air out of my tire. However, Joan also reported that a specific woman, whom I will call Martha, was about to cause me harm, and within three hours of the email, I learned from a third party that Martha had indeed attempted to do so. The event in question happened in the context of a legal matter, and it was highly improbable that Joan and Martha had collaborated.

All told, it seemed reasonable to conclude that at least something psychic was transpiring. However, as reliable and dramatic as this evidence was, it was not by itself sufficient evidence to conclude that Susy was actually watching over me and sometimes giving me useful warnings.

It was possible, as discussed previously, that Joan was merely being very psychic, engaged in remote viewing as well as precognition—the ability to foresee future events. The fact that Joan did not interpret it as remote viewing was not, in and of itself, justification to conclude it was anything more than this.

However, what happened next in this particular self-science venue was beyond my wildest imagination. In fact, it ultimately became a transformative discovery, provided by, of all people the deceased but spiritually alive Susy Smith, a spirit interloper par excellence.

Spirit's Backseat Driver

It was the fall of 2001. I was on the East Coast for a combined business and pleasure trip. (To preserve the anonymity of the people involved, I will keep the location vague.) On Saturday, I was being driven from the city to a suburb to visit with a grieving elderly couple whose adult daughter had recently died of brain cancer. Both the husband and wife were Holocaust survivors. On the trip, the elder daughter, whom I will call Alice, shared with me some of her younger sister's history.

Alice wished she could have a convincing conversation with a research-validated medium to assure her that her younger sister's essence was alive and well. Alice wanted her parents to know that contemporary scientific research was supportive of their hope that there was life beyond death and their wish that they would someday be with their deceased daughter.

As I listened to Alice, I realized that I had been having a truly unique personal scientific experience with Joan. I had been receiving the kind of information that, while it would not convince a skeptical scientist, would provide profound comfort to a layperson seeking evidence that their deceased loved one was well.

The truth is, I began to feel guilty. I deeply wished that Alice and her family could have the same comforting experience I was having about the possibility of life after death.

The next day, Sunday, I was back in the city and checked my email and could not believe what I was reading. There was a surprise email from Joan containing a seemingly incredible story.

Joan began by saying that she knew this was the weekend, however, something strange happened, even for her, which required that she write me at this time. She claimed that on Saturday morning, around the time that I was being driven to meet the elderly couple— of course, I had not told Joan my travel plans or what I was doing there—she was driving in her new car. Suddenly, Susy and an unknown deceased person appeared in her car.

I presumed that it must be distracting, if not dangerous, to be driving a car with deceased people as passengers. Joan wisely pulled over, communicated with Susy, and ultimately did a reading with the unknown woman. She further claimed that Susy instructed her to email the information from the reading to me immediately, supposedly insisting that I would know what to do with it.

I wondered: What if while I was feeling guilty in the car on Saturday and thinking about Susy and our watching-over experiment, Susy heard me and wanted to help this family?

What if Alice's deceased sister had been in the car when Alice was talking about her, and Susy had met the deceased woman?

And what if upon hearing Alice's as well as my private thoughts and wishes, Susy had realized she could barge in on Joan and ask her to do an impromptu reading with the sister for this family?

Three clearly speculative what-ifs.

As I looked over the information provided by Joan in the email, I could not tell if any of it was accurate except for a few details.

Remember that this was Sunday morning. I had a novel dilemma—should I call Alice, tell her what transpired, and then determine whether the information fit her deceased sister? Would Alice think that I was crazy? Or would she relish the opportunity to receive this unanticipated reading?

I ultimately decided that the only way to answer the questions, as well as to honor the apparent wishes of Susy, Joan, and the purported deceased sister, was to muster my courage and call Alice. I placed the call Sunday afternoon. I explained that something unusual if not downright weird had happened, and explained the circumstances. I asked Alice if she would be interested in hearing the email and scoring each item in a careful fashion. She said yes, and we spent an hour reviewing the information.

To my astonishment, Alice's scoring of the information indicated that the accuracy was greater than 80 percent.

However, as mentioned earlier, survival is in the details. One unique piece of information was the tipping point for this work.

Midway through the communication, Joan claimed that the deceased was saying that eagles were important to her. That I needed to tell the family about the eagles, and that they would understand what this meant.

Meanwhile, on the other end of the phone, Alice was sobbing. Remember, Joan did not know that I was meeting with Alice and her family or even that I had traveled to the East Coast. It turned out that the deceased sister indeed had had a passion for eagles. She collected statues and paintings of them. The eagle was like a totem animal to her.

Apparently, after she died and was cremated, instead of her ashes being placed on the front table at the service, the family set out a carved glass statue of an eagle. The song "Fly Like an Eagle" played at her memorial service. And to verify all this, the family later sent me a videotape of the service!

While Alice was explaining this to me on the phone, the thought crossed my mind that Susy had revealed a potential new proof-of-concept research paradigm. This surprise experiment conducted by the other side was suggesting that one deceased person could intentionally bring a second deceased person to a medium. In other words, the first deceased person could be, so to speak, serving as an afterlife, or spirit, experimenter!

Notice that using this new paradigm, it becomes possible to go beyond conventional double-blind experiments, and the success in the study requires active collaboration from the other side. Not only does this paradigm rule out the medium reading the mind of the sitter or experimenter, but it also implies that the medium is not simply reading dead information in space, since the accuracy of the information requires the intentional cooperation of a spirit co-investigator. This would offer further proof of discarnate intention—spirits acting with the same willfulness as the living—which helps establish spirits' presence and at least their ability to offer advice and guidance.

Possible future experimental designs rapidly flashed through my mind. Of course, I could not share these burgeoning plans with Alice; she was immersed in the emotional pain and comfort of discovering that the information fit her beloved sister, and it implied that her sister had cooperated with another deceased person to convince an unsuspecting medium to perform an unplanned reading to convince her family that she was alive on the other side.

After Alice and I hung up, I was both inspired and perplexed. I was inspired because what I had just witnessed went far beyond predictions from remote viewing or even superpsi. What I had witnessed was potentially revolutionary, proof-of-concept evidence of spirit-directed research, conceived and implemented by someone on the other side.

I was well aware that before I would propose formal university-based research using a double-deceased paradigm, which would indicate my belief that what I had experienced was a legitimate protocol, I would need to see it privately replicated many times. As it would turn out, what I had witnessed was metaphorically like the early-morning hint of yellow light beginning to rise above eastern mountains. In time, this streak would get bigger, brighter, and ultimately round out. The light would become so intense that you could only look at the glowing yellow ball through a pair of strongly filtered lenses.

Exploring the Double-Deceased Paradigm

Over the course of the next six months, I seized upon every opportunity that occurred spontaneously in my personal life to determine whether Susy's double-deceased paradigm would replicate. All of the self-science investigations proved to be successful. I will briefly share one of them here.

I decided to test the double-deceased paradigm with Mary Occhino. Mary is an extraordinary medium who hosts the highly successful daily satellite radio show *Angels on Call*. I've known Mary since 2003 and have been able to verify the accuracy of her mediumship on several occasions. She is featured in part IV of this book addressing the question of whether angels are real and play a protective role in our lives.

At first we tested the paradigm informally. This personal test with Mary involved the middle-aged granddaughter of a deceased medium. It is important that I keep all their identities anonymous. I shared with the granddaughter the emerging double-deceased paradigm research. It turned out that when Susy and the medium in question were both alive, they had known each other.

I suggested to the granddaughter that I would email Mary, inviting her to participate in the informal private experiment. I would ask her to contact Susy, and would explain to Mary that Susy was to bring along a second unknown deceased person, and I would request that Mary read the second person and provide the information in an email.

The morning of the reading, I called Mary to make sure she had received my email. She told me that while we were speaking, an elderly woman showed up, and Mary insisted upon describing her. She asked me if this sounded like the woman Susy would bring along, and I told her I was not at liberty to confirm or disconfirm this possibility.

However, after we hung up, I questioned whether this was the deceased medium, and I briefly mentioned what had happened to the lady I was dating at the time. She said that the woman Mary had described sounded like her beloved deceased grandmother! She was about the same age as the granddaughter of the medium whom Susy was supposed to bring to Mary.

The question popped into my mind: was I being gifted with two deceased grandmothers for the price of one?

I called Mary back and asked if she would do two readings. One with the deceased woman who had spontaneously shown up and the other with the woman Susy was supposedly going to bring. Mary gracefully complied. She completed the two readings by the early afternoon. Recall that Mary did these readings without knowing the identities of either of the granddaughters or their deceased grand-mothers. And recall that while I knew a little about the deceased medium, I knew nothing about the other grandmother.

Later that afternoon, I brought the two granddaughters together and asked them each to rate both sets of readings. The granddaughters identified correctly which reading belonged to which grandmother. Keeping in mind her mother's unique traits, each granddaughter's score-card for her grandmother was approximately 80 percent accurate while the other grandmother's reading was only 40 percent the same.

Though I am not at liberty to disclose details about these read-ings, the patterns of specific information obtained through Mary were so dramatic that blind judging would have produced exactly the same results.

This surprising, informal, private test of the double-deceased paradigm was unique because one of the deceased had seemingly shown up spontaneously and I had known nothing about her.

Think about this: if the double-deceased paradigm is legitimate, then the information should more accurately match the deceased person that the Spirit experimenter is allegedly bringing to the reading for the *targeted sitter*, the one expecting her specified loved one. This is in comparison with other deceased people who might show up, on their own, looking for their own relative.

The Double-Blind Test of the Double-Deceased Paradigm

Based on informal personal tests inspired by Susy Smith, I designed a formal proof-of-concept test extending the double-deceased paradigm to two experimenters and spirits. We would also have two physical experimenters like me and several sitters (those receiving the readings), all under double-blind scoring conditions—the mediums and sitters were not in contact. I did this as a private investigation with the collaboration of a scientist who was also privately exploring whether mediums could bring through survival evidence of his deceased daughter.

The two mediums who participated in the investigation were Joan and Mary. There were two physical experimenters: me in Tucson and the other scientist who was on the East Coast and whom I will call Dr. Ortega. The two deceased spirit experimenters were Susy Smith and Elizabeth Ortega—the name I will use for Dr. Ortega's beloved deceased child.

There were five professional sitters, located in different parts of the United States. All were professional and personal friends. Two were women: a physician and a grief counselor; three were men: a physician, a social worker, and the president of a small foundation. Susy was read by Medium Joan; Elizabeth was read by Medium Mary.

In the first phase of the investigation, Susy and Elizabeth each watched over one of the sitters on a given day. Joan would contact Susy for a reading, and Mary would contact Elizabeth for a reading. They would send their readings by email to a third experimenter, who coded the readings and then sent them to each of the five sitters for blind scoring to see if any of the information was related to them. Over the course of five days, each sitter would be watched over twice, once by Elizabeth and once by Susy. Using five envelopes in random order, I would open a given envelope on a particular day and ask Susy to visit the sitter named.

Dr. Ortega had a set of matching envelopes whose order was also randomized. He would open a given envelope on a particular day and ask Elizabeth to visit the sitter named in that envelope. Dr. Ortega kept me blind to the precise order of visits established by his envelopes, and I kept Dr. Ortega blind to the precise order of visits established by mine. Since the sitters were blind to which days they were actually being watched over by which deceased spirit experimenters, they did not know which two of the ten readings happened to be theirs versus the eight readings that belonged to the other four sitters.

The second phase of the investigation expressly involved the double-deceased paradigm. Spirit experimenters Susy and Elizabeth were instructed to each bring a deceased person related to a given sitter, which was again selected by randomized envelopes, to Joan or Mary.

The blind scoring of the readings showed that the spirit-assisted double-deceased paradigm was feasible as an experimental protocol. Statistically significant effects were obtained for the female sitters across both phases of the investigation—watching over and double-deceased—and for both spirit experimenters and mediums. The results for the male sitters were in the predicted direction, but they did not reach statistical significance because one of the male sitters happened to be read poorly by both mediums. In other words, his inability to be read was independently replicated by Joan and Mary. Interestingly, he was the one sitter of the five whose loved ones were long deceased, and he no longer felt an emotional need to connect with them. Minus his poor scores, the scores for the other two male sitters were similarly positive to their female counterparts.

This chapter reveals the possibility for establishing, once and for all, that survival of consciousness (such as intentional, living consciousness) is real and that our deceased loved ones can continue to be with us—and even educate, help, and guide us—if we are willing to listen. In other words, this can scientifically prove that the Sacred Promise is indeed a solemn covenant between Spirit and us.

As profound as the double-deceased paradigm is, it is just one bright star in a night sky filled with many sparkling stars waiting to be revealed.

We are about to consider another spirit-assisted experimental paradigm, revealed again by Susy Smith. The discovery of this paradigm included a celebrity, the late Princess Diana. It appears that well-known stars like Princess Diana are at least as persistent in being heard as are unknown stars like Susy, Shirley, Marcia, and Elizabeth. Some of them apparently intend to assist in future research and even help heal the world—and we can use all the help we can get.

6

PRINCESS DIANA SHOWS SPIRIT'S INTENT TO GUIDE AND PROTECT US

*I'm aware that people I have loved and who have died
are in the spirit world looking after me.*

—Princess Diana

If survival of consciousness is real, and therefore discarnate intention—including the genuine capacity for intelligence, creativity, choice, playfulness, and willfulness—exists after we die, then a major premise of many spiritual philosophies is plausible—that life after death exists. It is often claimed by spiritual leaders, as well as laypeople, that when we are in the physical, we are not only watched over by spirits but sometimes protected by them as well. And their protection can show up in the most surprising and even convoluted ways, as this chapter demonstrates.

In the previous chapter, I mentioned two examples from my personal investigations with Joan in which Susy purportedly gave me predictive messages that were warnings about the future—a flat tire and a potentially harmful act by a malicious woman. Since I was focused on the watching-over statements in that personal investigation, not the predictive and potentially protective ones, it is possible that Susy may have felt that I was ignoring an important aspect of her loving and caring concern for me.

Try to imagine that if you were deceased, and you were trying to offer guidance and protection to your loved ones, and they actually received the information but did not take it seriously; how would you feel? Remember that Susy was not only loving and tough, but she had a personal mission to prove that she is still here. I sometimes joke that Susy has given new meaning to the phrase, our work is never done.

The following account describes a completely surprising and bizarre occurrence during a carefully designed, formal, university IRB–approved experiment in which Susy taught me another profound lesson about life after death: the capacity of Spirit to protect us by choosing what information it will, and will not, provide in a reading. Moreover, the account in its entirety illustrates the creative cooperation that can apparently exist on the other side when sophisticated deceased people come together to not only participate in research but also teach the experimenters important lessons about the reality of the afterlife and some spirits' continuing intent to help us in our lives. As stated previously, in a deep sense, survival is in the details. What you are about to read may sound like spiritual science fiction, but it is all true, though certain names and minor details were changed for the sake of anonymity.

It appears that spirits can not only withhold key information, they can even potentially lie for the sake of protecting us, as in this instance. And they can intentionally mess up carefully designed university-based research for a higher purpose—and in the process, reveal a new research paradigm for documenting life after death and spiritual assistance in our lives.

Testing a Medium's Psychic Ability

In the spring of 2005, I was contacted by an American film producer who was doing a documentary about a European medium. I will call the medium Sandra. The producer asked me if I would be willing to test Sandra on-camera. The producer claimed that Sandra was extraordinarily gifted, and that she would provide me with important research data.

I had never formally tested a European medium. The woman would be in Connecticut at the time of the experiment. The producer would pay for the research, and the testing would occur in the summer—a lovely time to take a vacation, return to my former home in Guilford, Connecticut, revisit my old teaching grounds of Yale University, and explore some of Connecticut's beautiful coastal towns.

Also, the filming of the documentary provided me with the opportunity to conduct a unique kind of afterlife experiment over three locations: Arizona, Connecticut, and Europe. My idea was that the double-blind portion of the experiment would be conducted by Dr. Julie Beischel—at that time the William James Postdoctoral Fellow working under my direction—over the telephone in Tucson. The medium, producer, director, camera crew, and I would be in Connecticut, and the two secret research sitters would be in Europe.

Dr. Beischel and I added this medium to our by-then ongoing university-based double-blind mediumship experiment. In this double-blind portion of the experiment—where there is no contact between medium and sitter—Dr. Beischel in Arizona would ask the medium in Connecticut, over the phone, a set of standardized questions about the sitters' deceased loved ones. The sitters in Europe would not be on the telephone and therefore not hear the readings.

The information from the readings would be transcribed, emailed to the sitters, and scored blindly by them—meaning they would not be told which of the two readings was theirs.

In the single-blind portion of the experiment conducted by me in Connecticut, the medium would be allowed to speak with the sitters in Europe over the telephone. In both the double- and single-blind portions of the experiments, the medium was blind to the identities of the sitters and their deceased loved ones.

The testing was to take place over one weekend. For a given sitter and their deceased loved ones, the double-blind portion of the experiment would be conducted in the morning, the single-blind portion in the afternoon. Though this university-based experiment was novel in some ways, it did not include specific techniques that might reveal survival of consciousness. It was testing mediumship capability, not life after death.

For example, I did not employ the double-deceased paradigm because this would be too difficult to explain to the audiences who would eventually view the documentary. Hence, individuals like the late Susy Smith were not invited to participate in the research, given that we would not need a spirit go-getter. Let me reiterate: Susy was *not* supposed to be part of this experiment.

If positive results were obtained in this testing, the experimental design did not rule out the medium's possible mind reading of the experimenters and sitters or the reading of the information in the vacuum of space. The testing was to determine if Sandra was psychic, not to establish that she was actually receiving intentional communication from Spirit.

However, as you will see, the experiment that we had designed did not go as planned. Complications arose that ultimately were discovered to be spirit-initiated and illustrative of a novel and powerful proof-of-concept paradigm. Though everything you are about to read really occurred, I still find what happened hard to believe.

Princess Diana Becomes Part of the Experiment

The producer agreed that I would keep the identity of the two sitters and their deceased loved ones secret from the producer and her staff as well as from the medium. I chose two sitters, both female, who were in Europe at the time of the testing.

I will call the first sitter Mrs. Parker. The person she wished to hear from, her deceased husband, I will call Professor Parker. He had been a well-known psychology professor and scientist working in the Netherlands (those details are disguised). I had the privilege to know Professor Parker personally; he was a hero of mine. Respecting the wishes of his family, I have protected their anonymity. The second sitter was Hazel Courteney, a distinguished British journalist and author of numerous articles and books in the field of alternative medicine. She further reported on her experience in this experiment in her book *Evidence for the Sixth Sense*.

The person she wished to hear from was Princess Diana. Hazel had known Princess Diana and, subsequently, had spiritual experiences involving the late princess that I verified in mediumship research. Hazel had been deathly ill, and she felt that Princess Diana, in spirit, played both a protective and healing role in her return to health. Princess Diana's assistance supposedly ranged from recommending treatments to making recovery predictions that Hazel confirmed.

For both sitters, there was the possibility that the medium might receive information that the families would not wish to be made public. In all research involving humans, subject anonymity must be protected. This requirement was more severe in the present instance due to the levels of visibility of the deceased and their families, as well as the fact that the research was being conducted as part of a documentary film meant to be aired on television.

Following university and federal guidelines, I negotiated a written agreement with the producer that all filming involving research with the medium would be subject to human protection, and that I would have final say over what content from the research filming was included in the documentary.

Armed with this legal agreement, I could then show each of the respective sitters what information the producer wished to air, and each sitter could personally choose what aspects she felt comfortable in revealing. However, as we would all learn later, I did not have control over the B-roll or background footage when the formal research was not being conducted. Producers are entitled to control over their works of art, and rightfully so—presuming that their subjects have signed legal agreements giving them this control.

The two sitters were informed that the research was being conducted in the context of a documentary. However, they knew that the research information was being protected via the university-approved Human Subjects Consent forms that they had happily signed. As it turned out, the legal protections carefully put in place were insufficient to fully protect the rights of the sitters and the families of the deceased they represented because of what would transpire during the B-roll filming. I was not aware of this shortcoming—I cannot see the future—but apparently my spirit collaborators were, and they intended to do something about it. And interestingly enough, the revelation of this personal and legal complication was made apparent by, of all people, Susy Smith!

Sorry, Her Deceased Husband Is Sleeping

I arrived in Connecticut on Friday night. I briefly met the medium, Sandra, the producer, and the crew. Sandra was clearly nervous. She had never been formally tested by a scientist, and this experience was compounded by the fact that she was being evaluated as part of a documentary about her and her work. We reviewed the testing procedures that were to begin on Saturday morning. They understood that the identity of the sitters and deceased had to be kept secret from them. They did not know that the sitters and deceased were European. Our meeting was videotaped.

On Saturday morning, once the cameras were set up, I called Dr. Beischel in Arizona, she was put on speakerphone, and she began the experiment. Dr. Beischel followed the standard protocol we were using at the time. After introducing herself, she asked that Sandra receive whatever information she could about the deceased loved ones associated with absent sitter 1. Sandra was not told the name of the sitter or the names of the deceased. This was the least focused and most general of instructions in the protocol. As such, anyone could come through.

Sandra began talking. She claimed that she was seeing a woman, whom she described, and shared information associated with her. Since I knew little about most of Mrs. Parker's deceased family history—just something about Professor Parker—I had no idea who the purported deceased woman might be or even if Sandra was getting anything valid from her.

After about ten minutes, Dr. Beischel asked the second, more focused question. This time she told the medium that the sitter was interested in hearing from a specific person, the sitter's deceased husband, who used to call her by the secret nickname "Honeybunch" (the Parker family requested a substitute nickname be used in this book). Our experience is that when the request is more focused, the medium will often provide information more directly related to the deceased person the sitter wishes to contact.

What happened next was completely and utterly bizarre. The medium claimed that the female spirit was saying that the deceased man was upstairs sleeping, and that she was not going to wake him!

What? A deceased spirit was interfering with a reading? A deceased woman was claiming that the desired communicant was sleeping? Talk about alleged discarnate intention!

My first thought was that Sandra was a fraud, that she could not pass our rigorous research test, and that this was a clever ploy on her part to displace her failure and blame it on Spirit. I had never witnessed anything like this incident in over a decade of doing such research. Dr. Beischel, who at that time had only a couple of years in this field—her PhD was in toxicology and pharmacology, not in cognitive psychology and parapsychology—was speechless.

It did not make sense to continue this phase of the experiment since it would get even more focused: What does the deceased look like? How did the deceased die? What were some of the deceased's hobbies?

Before Dr. Beischel hung up, I asked Sandra, on camera, to converse with the deceased woman and see if she was willing to wake the gentleman later, thinking ahead to the single-blind portion that I would conduct in the afternoon. Purportedly the deceased woman said maybe.

Maybe? I was dumbfounded as well as highly skeptical of what was transpiring. The formal portion of the research reading ended. Dr. Beischel hung up thoroughly confused.

Sandra was upset; she felt she had failed the first portion of the mediumship test. The cameras kept rolling for the B-roll as I spoke with Sandra about her experience. What happened next shifted from merely bizarre to frankly alarming. Sandra began getting information about the sitter and her relationship with the deceased male, and I recognized that the emerging information was accurate. Moreover, some of what Sandra was saying was very personal and I seriously doubted Mrs. Parker would want it in a documentary film.

I could not sense if Sandra was getting this psychically—by reading my mind, by remote viewing—or through her guides. What was clear was that it did not sound like it was coming from the deceased Professor Parker. Sandra was not talking as if she were communicating with the deceased. She was not saying, "The deceased is showing me X" or "The man is telling me Y."

I quickly realized that I did not have control over this B-roll information. The producer and director appeared delighted with what was unfolding and the sensational information being revealed, and this concerned me greatly. This was not part of the formal research paradigm covered by the university Human Consent form, and it would potentially make for juicy viewing, and I was not in a position to protect Mrs. Parker by giving her the option of whether she wished to have this spontaneous information included in the documentary.

I faced an ethical challenge and immediately knew what I had to do. I told the producer, on camera, that some of the information Sandra was now getting was potentially related to the sitter and her deceased husband, and that I seriously questioned whether the sitter would want it aired. I told the producer that we would need to amend what I had signed to include any B-roll footage related to the research, or I would have to stop the experiment and withdraw from the project.

The producer was frankly furious, but with reluctance and begrudging acknowledgment, she agreed to amend our signed document accordingly. We decided to take a break and meet in an hour to add the necessary sentences. When this was completed, I would agree to conduct the afternoon, single-blind portion of the experiment. Meanwhile, Sandra was stressed out and exhausted. As it turned out, this medium was an innocent participant in all these complications.

I called Dr. Beischel, my co-experimenter. I brought her up to date about what had transpired and asked if she recognized who the deceased woman might be. On a few occasions, Dr. Beischel's deceased mother had spontaneously shown up in her daughter's experiments. We reviewed the characteristics of this deceased interloper as revealed by Sandra's reading, and quickly realized it did not fit her mother.

However, it dawned on me that at least 80 percent of the information—which included character traits—was consistent with someone who sometimes showed up unannounced in my research: Susy Smith.

And then it hit me: Susy had met Professor Parker when she and he were alive in the physical; Susy had admired Professor Parker and his work. She would have cared about him and his widow in addition to our research. And Susy would know that the medium was psychic and might pick up information about the deceased loved ones informally. Could Susy have intervened, possibly in agreement with Professor Parker, to ensure that Mrs. Parker, my laboratory, and our research were protected? Was I witnessing a potentially new proof-of-concept possibility? The question was both logical and reasonable.

Of course, there was no way I could know for sure, especially at that time, whether Susy had dropped in to the experiment for the express purpose of protection. Remember that this was not part of any specific protocol for the experiment or its experimental design.

But given Susy's history with me both before and after she died—as well as the discovery of the double-deceased paradigm—the evidence strongly supported the hypothesis that Susy was one creative, intelligent, caring, persistent, and tough spirit interloper. Little did I realize that this was merely a preamble to even greater surprises and lessons that would unfold over the course of that life-affirming and transforming weekend.

That afternoon I oversaw the single-blind telephone reading between Sandra and Mrs. Parker. Since the medium was now allowed to speak with Mrs. Parker, she quickly recognized the sitter's foreign accent. According to Mrs. Parker, as well as my own knowledge of her husband, Sandra obtained meaningful and accurate information related to Professor Parker.

However, Sandra said some things purportedly from Professor Parker about specific concerns he had in regards to Mrs. Parker that were not only upsetting to her but were rejected by her. Trained in clinical psychology, I saw the potential accuracy of the medium's insights and appreciated Mrs. Parker's difficulty in accepting some of the loving guidance and protective concerns supposedly offered by her deceased husband.

The Medium's Surprising Revelation

Both Sandra and the producer were relieved when the phone interviewed ended. Sandra appeared to be genuinely psychic, even though the experimentally rigorous portion of the test—the morning's double-blind portion—had been a complete failure, at least in terms of getting any information related to the allegedly sleeping deceased husband.

We went upstairs for cocktails and dinner. The cameras continued to roll. At one point Sandra said to me, "You may not know this, but I have had some very famous clients. Someone I was especially close to was Princess Diana, and I am going to confess this in the documentary."

I could not believe what I was hearing.

Think about this: Here I had decided to have Hazel Courteney be the secret sitter 2 for Sunday, and Princess Diana was to be the secret deceased, and it turned out that the medium had a longstanding relationship with Princess Diana! I wondered, what was the probability that I would pick Princess Diana to be the secret deceased subject, and the medium in question would happen to have a secret and deep relationship with the deceased? Princess Diana was known to consult with psychics and mediums, but how many of these were out there? Only a few, I would guess.

Meanwhile, if Sandra knew Princess Diana intimately, and if Sandra was a genuine medium, then she would quickly recognize the deceased in the double-blind portion of the experiment, and it would no longer be double-blind! In other words, the double-blind portion would fail again, this time for a different reason.

Two failed experiments in two days? On the other hand, I realized that because I now had editorial control of any B-roll footage related to the research, I had the potential to protect any information that might come through related to Princess Diana. If ever a deceased person and her family deserved anonymity for selective information, it was her and her family.

I wondered to myself, would Susy be aware of this fact? Would Susy, along with Professor Parker, team up with Princess Diana to ensure that all parties were protected in this emerging new research paradigm, at least in terms of the carefully designed double-blind portion of the experiment?

Though I was relieved that I now could exert legal protection of what might transpire next, I was alarmed that Sunday's double-blind portion would be a flop.

Once again, however, I was wrong. What transpired was seemingly unbelievable and profoundly meaningful concerning the way Spirit can potentially protect and educate us.

The Unbelievable Double-Blind Reading with Princess Diana

On Sunday morning, I called Dr. Beischel, put her on speakerphone, and she conducted the double-blind session. She began by reintroducing herself and requested that Sandra receive whatever information she could about the deceased loved ones associated with absent sitter 2. As in Saturday's double-blind session, Sandra was not told the name of the sitter or the names of the deceased. Theoretically, anyone could show up.

Sandra said she saw a woman. The woman was fairly tall and slim and showed herself in the countryside. Curiously the woman did not show her face. Dr. Beischel then asked the second, more focused question. She told Sandra that the sitter was interested in hearing from a specific person, a female acquaintance of the sitter; we didn't say "Diana" for obvious reasons. Sandra continued to describe a mysterious woman who seemingly would not give her name or show her face. I thought to myself, isn't that interesting (and odd). But I once again wondered, was this a ploy of Sandra's, using Spirit as an excuse for her inability to obtain what for skilled mediums is relatively straightforward information? Or was something more going on?

Dr. Beischel then asked the specific question: "What does the woman look like?" In response, Sandra described a fairly tall and slim woman with relatively short and straight, light-colored blond-and-brownish hair. Sandra described her as wearing a simple, full-length dress, in the countryside near a typical-looking farmhouse. Again, Sandra claimed that she could not see the deceased woman's face.

In response to, "How did the woman die?" Sandra said that the death was relatively quick and that it involved an automobile accident.

In response to, "What were the woman's hobbies?" Sandra said that the woman showed herself and her sons skiing, and that she really cared about her sons.

As I listened to the information, I realized that the pattern could have easily fit Princess Diana, presuming that you knew who the deceased was. The specific information did not fit Dr. Beischel's mom or mine, or Susy Smith, for that matter. I was fully expecting that Sandra would say: "I recognize this woman. I knew her. This is Princess Diana." Instead, Sandra claimed that she had no idea who the deceased woman was!

After Dr. Beischel finished the formal part of the experiment, we entered the B-roll postexperimental portion. I pressed Sandra, asking her if she had any idea who the deceased was. She insisted that the woman was not very forthcoming and that she would not give her name or show her face. In fact, Sandra insisted that the woman was often whispering to her during the reading.

Whoever heard of a deceased person whispering? Sandra claimed to be as confused as I was. The more I pushed her, the more she resisted. I was forced to conclude that either Sandra was one of the most skilled actresses I had ever met, or she was really stymied and did not know who the mysterious female spirit was.

Of course, if Sandra really could not identify the deceased, then the double-blind portion of the experiment would be successful. Whereas my fear had been that the session would be confounded by Sandra's former relationship with Princess Diana and therefore be a failure, the double-blind portion of the experiment seemed to have worked.

Think about this. If Sandra had been reading my mind (or Dr. Beischel's), she would have known this was Princess Diana.

Then it dawned on me: if Susy and Professor Parker could intentionally withhold information during Saturday's double-blind session, could Princess Diana intentionally withhold her identity during Sunday's double-blind session as well?

Remember that Princess Diana has participated in a previous mediumship experiment with me related to Hazel Courteney. And although the idea that Princess Diana was purposefully withholding her identity was reasonable in light of what was unfolding, I had no idea as to its truth.

After the B-roll session was completed, we took an hour break. I drove my rental car back to the hotel and called Dr. Beischel. I told her what had transpired, and she agreed that the information received in the double-blind portion fit Princess Diana to a tee. And like me, she found it extraordinarily difficult to believe that Sandra did not recognize her purported friend Diana.

I returned for the afternoon single-blind telephone session more confused than ever. I called sitter 2, Hazel Courteney in London, turned on the speakerphone, and allowed Sandra to introduce herself. I presumed that Sandra immediately recognized that Hazel had a

British accent. Sandra then said, to my utter amazement, "I see this deceased woman, and she is turning around ... and ... Oh my God, it's Princess Diana!" At this point, Sandra broke down sobbing and relative bedlam ensued.

Though my heart went out to Sandra, my first thought was, "Oh great, now the single-blind experiment has been ruined! The medium has figured out who the deceased is, and moreover, she knew her intimately."

Hazel then confessed that she had wished to hear from Princess Diana. Whatever scientific value the single-blind session might have had was thoroughly ruined.

Meanwhile, through her tears, Sandra said, "Now I understand why the deceased was whispering to me. You see, on numerous occasions when she was alive in the physical, Princess Diana would call me for psychic and spiritual advice. I would bring through deceased relatives and spirits who cared for her. It was important to Diana that people not know that she was in contact with a psychic medium, so she would whisper to me on the phone."

Who would have guessed that a whispering spirit would turn out to be so meaningful and evidential?

Who would have thought that this little detail would be consistent with the hypothesis that not only does consciousness survive with intention but intelligent, creative, and caring spirits can take control over a carefully designed, university-based experiment and provide invaluable lessons for all of us—that is, if we are willing to listen. A new proof-of-concept research possibility was being born.

Proving that Spirits Have Control over Information

Think about this conundrum. Presuming that Sandra, Dr. Beischel, Hazel, and I were not engaged in fraud—and I can state categorically that we were not—the pattern of information revealed during this two-day testing period cannot plausibly be explained in terms of Sandra reading Dr. Beischel's mind or mine, let alone the two sitters'.

Moreover, the information, as it came to light, did not appear to be dead or retrieved from the vacuum of space by Sandra.

Instead, it looks like it was given to Sandra both for protective and educational purposes. Susy Smith, Professor Parker, and Princess Diana were each playing a part in demonstrating that they were there and in charge, at least of their own continued lives, and that they cared about their loved ones as well as afterlife research.

If you find this hard to believe, welcome to the club. If I have learned any lesson in participating in this research, it is to be prepared for surprises.

We began this chapter with a question: can Spirit protect the living? I end this chapter with another question: can Spirit manage its own information?

Can Spirit choose to give or withhold specific information? If it can, and we can continue to document this proof-of-concept possibility in future controlled research, does this help establish the fundamental premise that survival of consciousness is real and that Spirit can choose how it will interact with us in our daily and collective lives?

Though the documentary was apparently never completed, the producer was correct. Sandra did indeed provide me with important research data, and far beyond what I could have ever imagined.

And if you can believe this, the journey is just beginning, and more shining stars are about to appear.

PART III

The Promise of Spirit-Assisted Healing

7

CAN SPIRIT PLAY A ROLE
IN PHYSICAL HEALING?

Illness can offer opportunities to tap into
unseen powers beyond the physical world.
When we reach out, the world of the spirit becomes ours.

—Allan Hamilton, MD

When I was a child, I experienced a number of Passover Seders, which celebrate the Old Testament account of Moses and his role in helping free the Jewish people from slavery in Egypt. The story typically includes numerous supernatural elements, including the claim that God miraculously parted the Red Sea, thereby allowing the Jews to flee and drowning the pursuing Egyptian army.

Though I sympathize with the bondage of the Jews, as I do for any peoples cruelly enslaved against their will, I find it impossible to believe that a gigantic body of water was intentionally and mysteriously separated by an invisible force. As far as I know, there is no historical evidence that such a thing happened, nor is there scientific evidence that such a thing is possible. (I have recently heard, though, that some scientists have speculated about how, under very special wind circumstances, relatively shallow water covering a high ridge in a sea floor might be temporarily blown aside.)

I presumed that beliefs about spiritual healing were like beliefs about the parting of the Red Sea. I had been educated to believe that spiritual healing was either a myth, a lie, or the result of mind-body placebo effects.

However, the truth is that if the deceased and higher spiritual beings actually exist then it becomes possible, in theory, that they might be able to play a guiding, if not healing, role in the treatment of physical and mental diseases. And I might add, its opposite is equally possible. In South America, legitimate mental health practitioners consider psychiatric disease to be spirit possession, and they've had success curing people after exorcizing the evil spirits. I know that some might consider this type of diagnosis a stretch, but there is supporting evidence.

So, if Susy, Shirley, Marcia, Elizabeth, and Princess Diana still exist as explored in part II, then not only could Einstein still exist but so could gifted deceased surgeons, nurses, and other healing professionals. And, if they are eager to help us find lost documents, as in Marcia's case, just think of the motivation to help sick loved ones.

In light of the observations reported in part II, it no longer becomes such a stretch to justify the hypothesis that spirit guides could potentially play a role in healing, either in assisting healthcare professionals on earth or working directly with all of us in the context of self-healing.

For both scientific and personal reasons, I was led to face this possibility head on. In this chapter, I describe two real, life-transforming examples. As I share them, try to imagine what it would have been like if you had been in my shoes, (1) seeing a distinguished psychologist push an ice pick through his cheek painlessly and with no subsequent evidence of a wound or (2) experiencing your severe back pain virtually disappear after inviting alleged spirits to help remove your pain. I invite you to wonder what it would be like if you had witnessed and experienced such startling events. What conclusions would you have reached?

Be forewarned, what you are about to read contains claims and information that are unavoidably bizarre and challenging to the conventional Western mind. I share them because they really happened and because they place in context the overarching question of whether Spirit can and is willing to play a role in healing and how we can document it scientifically.

A Demonstration of Deliberately Caused Bodily Damage and Rapid Sufi Healing

In the winter of 2002, I was giving an address at an International Qigong Association meeting in San Francisco, and I met a distinguished clinical psychologist who shared a seemingly unbelievable story about himself.

To place his story in context, let me provide a bit of his background. His name is Howard Hall. He holds two doctoral degrees: a PhD in experimental psychology from Princeton University and a PsyD in clinical psychology from Rutgers University. Dr. Hall is a tenured senior faculty member at Case Western Reserve University in Cleveland, where he practices behavioral medicine with families, especially children suffering with chronic diseases. He has a background in behavior therapy, hypnosis, and biofeedback. His wife is also a healthcare professional.

Dr. Hall is also a deeply spiritual person, a student and adherent of the philosophy and practice of Sufism. This offshoot of the Islamic faith adheres to a series of mystical beliefs and practices. Sufism is to Islam as Hasidim is to Judaism. Most people are familiar with the Sufi whirling dervishes, which is actually a form of active meditation.

One of the controversial practices of a Sufi school in Iraq called Tariqa Casnazaniyyah (roughly translated: "the way of the secret that is known to no one") is referred to as Deliberately Caused Bodily Damage or DCBD. According to Dr. Hall, this refers to feats where dervishes intentionally cause serious damage to their bodies, yet with complete control over bleeding and infection, and unusually fast wound healing.

It is claimed that this practice includes but is not limited to hammering daggers into the skull, insertion of steel spikes and skewers in the body, chewing and swallowing glass and razor blades, and subjecting oneself to fire and to venomous snakes. It is further claimed that it involves control over bleeding and pain, wounds healing within four to ten seconds, and no harm to the body.

The practitioners believe that these effects are made possible because of the active assistance of their deceased Sufi ancestors in combination with Allah—their name for the Sacred. Skeptics presume that the practitioners must be faking the apparent insults to their bodies. There is little question that fakers exist but, as with firewalking or yogi siddhas or Christian miracles, fakers shouldn't dissuade one from considering legitimate demonstrations of the same.

Until I met Dr. Hall, and subsequently participated in an exploratory demonstration investigation of the phenomenon myself, I too presumed that the claims of DCBD were probably bogus. As it turned out, my presumption was wrong; some claims are literally as real as steel knives.

This practice is not unique to Sufism. The Sun Dance of Native Americans may include skewers through the skin, and Hindu devotees sometimes pierce their bodies with needles, hooks, and skewers.

Indian mediums are known to beat themselves with swords, and in Sri Lanka, people sometimes hang from hooks through their bodies. However, certain members of the Tariqa Casnazaniyyah school apparently carry the practice of DCBD to the extreme.

For numerous ethical as well as practical reasons, this practice has not been systematically studied in research laboratories in the West. (Paramann Programme Laboratories in Amman, Jordan, has conducted DCBD research with these dervishes.) The most obvious reason involves potential risks to human subjects—let alone possible legal risks to universities as well. Who would volunteer to collaborate in such studies? Since it involves spirits protecting someone from severe bodily injury, though, it falls within my research interests.

A precursor to my investigation was research by Eric Peper, PhD, and his colleagues, who studied a sixty-three-year-old Japanese yogi with thirty-seven years of experience and practice with spirit protection. The yogi would pierce his tongue with a metal skewer but reported no pain, and purportedly showed no evidence of bleeding or physical injury such as scaring or inflammation.

Dr. Peper and colleagues recorded EEG brainwaves before, during, and after the tongue piercing. They reported that during piercing, the yogi showed increases in low-frequency EEG activity, primarily sub-delta, delta, and theta, a pattern of brain waves suggesting that the yogi had somehow turned off his brain to external as well as internal stimuli.

However, we do not know how the yogi's brain was turned off to the stimuli. Did he do this all by himself—that is, with his mind alone? Or did he receive assistance from ancestors as well as connecting with the Sacred? Yogic masters typically believe in the power of ancestors—the living assistance of deceased yogic masters—as well as in connecting with the Source itself. *Yoga* literally means "union."

Dr. Hall had heard of this mystical practice and wanted to witness the claims himself—little did he know he'd end up volunteering for a personal DCBD test. In November 1998, he traveled from Cleveland to Baghdad and took extensive videos and photos of seemingly impossible acts of self-inflicted bodily damage. After Dr. Hall had witnessed and videotaped an Iraqi sheik perform a full range of DCBD feats, the sheik suggested that Dr. Hall try it for himself. Since Dr. Hall regularly meditated and prayed in the Sufi tradition, he was open to this spirit-assisted healing. Following the sheik's invitation, but with significant trepidation, he allowed himself to enter the prayerful state and requested the assistance of the Sufi ancestors and the Source.

A metal skewer was then passed through Dr. Hall's cheek. To his astonishment, he felt virtually no pain. Moreover, when the skewer was removed, he showed virtually no bleeding and the wound healed in less than a minute, leaving no evidence of inflammation or scaring.

Though reluctant to press his luck, Dr. Hall realized the importance of replicating this apparent effect in a research laboratory. When *National Geographic* wanted to film a documentary on DCBD for television, I was invited to participate in an exploratory investigation of Dr. Hall while they videotaped him passing a metal skewer through his own cheek without damage to it. This investigation involved Dr. Hall testing himself while we measured the effects.

We decided to record nineteen channels of EEG brain waves before, during, and after Dr. Hall pierced his cheek. In addition, we decided to use a gas discharge visualization (GDV) electrophotography device developed by Dr. Konstantin Korotov, a physics professor at Saint Petersburg University in Russia. The GDV camera was designed to measure biophysical energy fields around the body. We made a set of GDV recordings before and after the skewering period.

The tools we used in my laboratory—EEG and the GDV device—have been approved for use by numerous university IRB committees, including the University of Arizona. But this was not a

University of Arizona research project. I agreed to allow Dr. Hall's personal demonstration to be filmed in my laboratory because (1) this exploratory demonstration investigation was being performed for a public documentary filmed and funded by *National Geographic* (they used standard human consent and release forms); (2) the subject was Dr. Hall, a university scientist and clinical professor himself; (3) Dr. Hall had requested that this consensual self-science demonstration be conducted in my laboratory; and (4) we were using previously approved biophysical recording instruments.

Lab personnel present at the investigation included an EEG technician, a research nurse, a research medium, and an observer-only PhD psychologist in addition to Dr. Hall and me. And we hoped there would be those present from the other side who didn't sign in in advance.

To ensure that no trickery was potentially involved, I purchased a metal ice pick from a local Tucson department store and kept it in my possession until it was time for Dr. Hall to skewer himself. I can confirm that my ice pick was indeed used by Dr. Hall during this experiment.

After pre-skewer GDV electrophotos were taken, Dr. Hall closed his eyes, and meditated and prayed for almost ninety minutes. EEGs were continuously recorded during this period. Dr. Hall signaled us when he was ready to attempt to insert the ice pick. I had never tried to pass an ice pick through my cheek, or anyone else's for that matter; apparently, it is not as easy to do as one would expect. Dr. Hall pushed and pushed for approximately a minute before the ice pick finally passed through his cheek. After it was in place, he kept his eyes closed for a five-minute period during which time we recorded his EEGs.

Upon completion of the five-minute postinsertion EEG period, we invited Dr. Hall to remove the ice pick. He did this relatively effortlessly without any verbal or physical reaction to the possible pain.

The research nurse and I immediately examined his cheek. A single drop of blood appeared approximately twenty seconds after the ice pick had been removed. The drop was blotted with cotton. After approximately sixty seconds, the wound was examined for evidence of inflammation or tearing. Neither the nurse nor I could see any evidence of a wound, period.

We then took post-skewering GDV measurements and interviewed Dr. Hall. He told us that he had been surprised and concerned that the ice pick was so difficult to pierce through his cheek. However, he claimed that he felt virtually no pain during the insertion or post-insertion period. I knew that if Dr. Hall had been lying, or deceiving himself, evidence of stress and pain would be apparent in his EEG readings. In a follow-up conversation, after Dr. Hall had returned to Cleveland, he said he experienced no evidence of inflammation, scaring, or infection in the days and weeks following the experiment.

We later performed analyses of three five-minute periods of the EEGs: the first five minutes of the meditation and prayer, the last five minutes of the meditation and prayer, and the five minutes when the ice pick was in his cheek. The meditation and prayer were unremarkable; he showed a prototypic eyes-closed relaxed brain, with evidence of lower frequencies—theta and delta—in the frontal regions of the brain.

However, the EEG frequencies during the ice pick period were notable. There was no evidence of the increased EEG frequency activity that one would expect if Dr. Hall had been in pain or stressed; moreover, there was no evidence of increased scalp muscle tension either. Instead, his brain showed lower EEG frequency activity—more theta and delta activity. His brain looked similar to the yogi's in Dr. Peper's experiment, who was presumably doing his version of connecting with Spirit.

When we analyzed the GDV measurements, the pre-skewering patterns were also unremarkable. The energy field indices displayed on the screen were relatively uniform across Dr. Hall's head and shoulders. However, the post-skewering patterns were notable. There was a dramatic absence of a measurable energy field on the left side of Dr. Hall's face in the region where the skewer had been inserted. This specific pattern of findings could not be explained as an artifact of the recording procedure.

Though Dr. Hall claimed that he had not been self-hypnotized, he had been in a relaxed state. Clearly, something had happened here.

It is evident that Dr. Hall believes in the philosophy of Sufism; could this set of findings be entirely due to mind-brain-body effects? In theory, yes. Dr. Hall's DCBD test on himself does not address the question of how DCBD occurs; it only demonstrates that it can occur under special conditions.

However, was it possible that Spirit was somehow involved in producing this striking recording, indicative of its assistance during the experiment? I invite you to try and imagine what you would have thought and felt if you had witnessed all of this personally. Moreover, I encourage you to imagine what you would have thought and felt if you had been in my shoes and had interviewed the research medium and the PhD psychologist who happened also to be a gifted medium, both present during the experiment.

They each claimed that they had seen numerous deceased spirits appear in the laboratory during the meditation and prayer and skewering periods. In fact, though I found this especially difficult to accept, each of them claimed that Mohammad himself had showed up as well. (This would be equivalent to Jesus making an appearance at a Christian revival.)

Since neither the research medium nor the psychologist was blind to the purpose of the experiment, it is obvious and essential that we treat their reported sightings with caution; it is possible that their experiences were clouded by their expectations. Because this was an exploratory investigation performed by Dr. Hall on himself, not a systematically controlled experiment, I did not consider their observations worthy of video recording or reporting in detail.

On the other hand, both of them had been tested numerous times in the laboratory, under blind conditions, and each had revealed striking and repeated evidence of psychic sensing. Though their sightings of spirits during this experiment are not definitive, they deserve our serious consideration—especially in light of the remarkable GDV measurements.

The important take-home point is that the combined results point to the need for future blind experiments to determine if Spirit can show up in such sessions and play a protective and healing role. Though Dr. Hall does not see dead people, nor does he claim to have relationships with angels—both manifestations being tenets of Sufism—he believes that what he experienced, and what we witnessed, was made possible by Spirit.

It is worth remembering that Dr. Hall is not a swami, a shaman, or a spiritual healer. He is a well-respected Princeton- and Rutgers-trained experimental and clinical psychologist who is a tenured professor at a major university. It is one thing to dismiss the claims and brain of a sixty-three-year-old Samrat yogi; it is another to dismiss the claims, training, and self-discipline of a highly educated and successful mainstream psychologist and scientist.

Although Dr. Hall did not know my personal life history in this regard, I had direct personal familiarity with spiritual healing and the purported therapeutic effects of ancestors and other spiritual beings. My personal proof-of-concept experience involved the rapid and dramatic removal of severe back pain, followed by the completely unexpected and immediate reestablishment of even more severe pain to prove a point.

Save for my closest colleagues and friends, up to now I have not shared this experience with anyone. I realized that with the decision to write this book, and ultimately this particular chapter, if there ever was a time and place to reveal this illuminating experience, it was here in this context. How can I expect anyone else to "walk their talk" if I do not do this myself?

As a general rule, findings from university laboratories, though important to science and knowledge, are typically not life-changing. However, evidence from the laboratory of one's personal life can often alter the course of your life in surprising and even compelling ways.

What you are about to read did not involve EEGs or GDV measurements. There were no cameramen or research nurses present to monitor the situation. It was just me doing a headlong dash into new experiences, such as furniture moving, in this case. However, the event was no less real or important.

How Do I know This Isn't the Placebo Effect?

About twenty-five years ago, when I was a professor at Yale, I did something that was really foolish. I had purchased a six-foot tall wooden grandfather clock and decided I would try to carry it down a small flight of stairs, by myself.

As I leaned over, holding the huge clock cabinet, I felt something snap in my back, followed by excruciating pain. I had great difficulty straightening up. A medical examination of my back suggested that I might have pulled a ligament and/or ruptured a disk. I was living in conservative Connecticut, affiliated with Yale's medical school, which viewed chiropractors and massage therapists as voodoo doctors at the time. I was encouraged to take muscle relaxants and baths and to put as little strain on my back as possible. For the first week or so, I spent most of the time in bed.

For almost three months, I kept my chair-sitting time to an absolute minimum. I set my computer on boxes at my home office workstation so I could type standing up. The pain was gruesome at first. However, with time, it subsided, and I did not require back surgery.

Though my back recovered, it was left weakened and wounded. Every six months or so, like clockwork (no pun intended), my back would go out, and I would need to give it rest. Sometimes this would happen because I did something stupid, like lifting a box of books. Other times it would go out in the process of doing little things, like bending over to put a single book away on a lower bookshelf. I have learned over the years that the best medicine for treating my back is significant bed rest and lying down on a couch—essentially giving it and the rest of my body a vacation. Typically, a few days of reclining-focused rest is sufficient to enable me to resume normal functioning.

It was January 1, 2000. The new millennium had begun uneventfully; the anticipated worldwide computer crash had not occurred. However, I had just made what I called my millennium resolution—a resolution for the next one thousand years. It was simple, yet profound. I made the decision that I would choose to live the rest of my life as if survival of consciousness were true.

At that time I was in the process of finishing the first draft of *The Afterlife Experiments*, so this hypothesis was very much on my mind. Let me repeat: I made the decision that I would choose to live the rest of my life as if survival of consciousness were true. The key words are *choose* and *as if*. Here was my reasoning:

On the one hand, if survival of consciousness did not exist, then when I died, I would lose everything, including my awareness. Therefore, since I no longer existed, I would never know that my choice had been mistaken. I would never know that I had been wrong to live my life as if survival of consciousness were true.

On the other hand, if survival of conscious did exist, then when I died, I would continue to be aware, to think, have memories, make choices, and so forth. Therefore, since I still existed, I would then know that I had made the right choice. I would also know that my preparing for the future beyond the body had been prudent and worthwhile.

Remember, this was a thought experiment, the kind Einstein used to do when he was in the physical.

I was not drawing a scientific conclusion prematurely about the truth or fiction of life after death. What I was doing was mindfully considering the options and consciously choosing to make certain decisions with the understanding that survival of consciousness might be true. Little did I realize that I would be immediately tested to see if I would choose to act on my potentially long-term resolution. Here is what happened.

On January 2, 2000, I was bending down to put a book on a lower shelf, and my back went out. Just like that, and I was in severe pain. It had been at least nine months since my last back episode. However, I really wanted to go to my laboratory at the university, so I decided to take a shower. I gingerly climbed into the tub, turned on the water, and experienced the pain associated with standing still.

It then occurred to me that I had not implemented my millennium resolution. By this time I had received training in various healing techniques, including healing touch, Reiki, and Johrei, and I was familiar with the claims that experienced healers often invited Spirit assistance in the form of deceased people, angels, and divine energy. I had never thought to ask for Spirit assistance for myself (or anyone else, for that matter).

There I was, standing in the shower, and I decided I would make the request, in my head.

I said to myself, "If anyone in Spirit can help my back, I would greatly appreciate it." I sent my request to various deceased beings that I knew of, including the deceased founder of Johrei and Sam—short for Shemuel, in Hebrew the name of God.

After about five minutes, as I got out of the shower and began drying myself, I was shocked to experience that my back pain had decreased a good 80 percent. On a scale from 0 to 10, with 0 being no pain and 10 being severe pain, my back pain had gone from an 8 or 9 to a 1 or 2.

I could not believe it. My back pain had decreased dramatically within minutes. This had never happened before.

After I had finished shaving and was I was getting dressed, I began considering possible alternative explanations for the abrupt and dramatic reduction in my pain. As I was putting on my pants, I innocently thought the following question: *How do I know that this is not a placebo response?* Remember, I am a scientist and questioner and this is an obviously fundamental question.

What I heard next completely took me by surprise. As I write these words, I can vividly see myself back in that bedroom, and it is as if the experience happened yesterday.

I quietly yet clearly heard in my head, "Very simply, we will take our support away."

Upon hearing the words "take our support away," I immediately experienced the most severe back pain I could remember—at least as gruesome as the original pain from the clock-moving disaster. On a scale from 0 to 10, it was at least a 12.

I fell onto to the bed, crumbled in a fetal-like position, with my pants half on. I had never experienced anything like that pain. It was so severe that I could finish neither pulling up my pants nor taking them off. I was virtually immobilized with pain.

Meanwhile, my intellectual—and often playful—mind had the following thought, *Now this could be a superplacebo effect!*

I realized that I had either:

1. Experienced a double-placebo response where:

    A. I had imagined that Spirit would help take away my pain, and then

    B. I had imagined that Spirit said if it would remove its support, that my pain would come back, in spades; or

2. I now understood—in a new and deep way—the larger meaning of the idea that it could have been so much worse.

If you are a skeptic and/or a conventional psychologist, you will presume that what I experienced was either a rare chance event and/or some sort of double-placebo (mind-body) effect. Either way, the explanation would have nothing to do with Spirit per se.

However, what if Spirit *was* involved in my rapid pain relief, followed by the even more rapid return of pain symptoms?

What if Spirit is actually assisting all of us, to various degrees, all of the time but we don't recognize it?

What if "it could have been so much worse" applies 24/7, because the normally invisible Spirit world is constantly providing us with background support and guidance?

What if it is in our power, if we so chose, to sincerely ask for additional Spirit assistance, and then actually receive it?

What if all of us have spirit guides, including access to Universal or Spiritual Healing Energies? (The latter leads to the acronym SHE, the Divine Feminine.)

Lying in bed, while doubled over in pain, I was not about to attempt solving this mystery at that time. However, I did realize that I was being given the opportunity to ponder this mystery and decided to give my back the reclining-focused time it seemingly needed.

I spent the next few days reading spiritually focused books I had never taken the time to explore, including *The Autobiography of a Yogi* as well as a book by the founder of Johrei, so that I would not lose awareness of the potential take-home spiritual message here.

When You Hear Hoofbeats, Don't Look for Spirit First

There is a well-known phrase in emergency medicine that I believe applies equally well to science: when you hear hoofbeats, look for horses, not zebras. The reasoning is as follows:

In emergency medicine, where life-altering decisions often need to be made quickly, it is prudent to consider the most probable causes for a given situation first, and then move on to the less probable. In other words, if the given situation or symptom is, metaphorically, hoofbeats, then the correct phrase would read something to the effect, "When you hear hoofbeats, look for horses first."

If the horse explanation was not supported by the facts, the emergency physician would then consider the next most probable response, which might be a pony or a mule. (Remember, this is a metaphor; the precise probabilities are not important here.) Clearly, it would not be time to look for a zebra explanation until horse, pony, mule, donkey, and maybe moose explanations had been sought and ruled out.

When we consider the two single-subject sets of observations reported in this chapter—Dr. Hall's apparent DCBD that he demonstrated in the university laboratory and my apparent back-pain reduction and then recurrence in the personal laboratory of my private life—the explanation for the hoofbeats we heard would appropriately be the horse (the placebo/mind-body explanation) rather than the zebra (the spirit-assisted explanation).

However, when we take into account the information presented in part II of this book, there is reason to entertain the hypothesis that Spirit exists. Consequently, it is prudent that we include the zebra—spirit-assisted effects—in our list of possible explanations. And it's now possible for science to test whether Spirit actually shows up in healing sessions. Spiritual healing can now be conducted under blind conditions, whether the person who is sick knows that they are receiving healings or not, as you will see in the next chapter.

8

TESTING SPIRIT'S PRESENCE IN HEALING SESSIONS

Intense love does not measure; it just gives.

—Mother Teresa

You will recall that Dr. Hall believes that spirits may have played a critical role in his ability to push an ice pick through his cheek in such a way that he could (1) experience virtually no pain, (2) have minimal bleeding, and (3) manifest almost immediate wound healing. A research medium and a PhD psychologist both claimed that they saw a collection of spirits present during our exploratory investigation with Dr. Hall.

As we discussed in the previous chapter, it is scientifically responsible to consider first various horse-like explanations—such as fraud,

117

misperceptions, and placebo or mind-body effects—before entertaining zebra-like explanations—such as genuine assistance from deceased physicians, angels, and ultimately, the Sacred.

However, a challenging question arises: is it possible to scientifically document whether Spirit was actually present in a given healing session? If we cannot address this fundamental question, it makes little sense to attempt to address the question of whether Spirit can play a guiding, if not mediating, role in the healing process. This chapter and the next present findings from two novel and exploratory proof-of-concept personal investigations conducted in my private life—and replicable in laboratory research—that addresses the question: can we document the presence of spirits in healing sessions?

Though both real-life investigations seemed to have occurred fortuitously and serendipitously, the healers and mediums involved claimed that a collection of spirits was intimately involved in orchestrating the opportunity for these most surprising discoveries. Moreover, the healers and mediums claimed that the effects we were documenting were actually about the giving of love.

A Healer's Claim of Spirit Assistance from a Deceased Physician

Shortly after my millennium resolution, I met an energy healer with a PhD in psychology who claimed enthusiastically that his grandmother, a deceased physician, often helped him in his healings. To retain their anonymity, I will call the healer Dr. Michaels and his grandmother Dr. Jones (a few other minor details are also disguised).

Dr. Michaels claimed that sometimes Dr. Jones showed him where to place his hands, and at other times she literally entered Dr. Michaels body and guided his hands.

Dr. Michaels loved his grandmother and deeply admired her. Though Dr. Jones was a Western-trained physician, she was a deeply spiritual person. Many of his grandmother's former patients were convinced that Dr. Jones's healings involved more than the medicines she prescribed and the caring treatment she provided. She apparently prayed for her patients and asked for Spirit's assistance in her patients' recoveries.

Dr. Michaels consistently found through clinical experience that if he called on his grandmother's help—in either his healings or his personal life—she would always assist him.

I asked Dr. Michaels how he knew that his grandmother was present. He said that he would sometimes see her at the foot of the massage table or would hear her, and at other times just feel her presence. Dr. Michaels said that some of his patients would report seeing a female spirit in the room, too.

Dr. Michaels claimed that he typically experienced his hands warming up and even becoming hot when Dr. Jones was ready to work with him. Sometimes his hands would get hot early in the session, just as he was beginning his energy diagnosis period, and at other times he would experience his hands getting hot later in the session, shortly before he was to begin his energy healing period. This particular claim caught my attention because I realized it was a procedure that could be potentially manipulated and measured.

My "disease called science" automatically kicked in: Dr. Michaels's belief—"I sometimes experience my hands getting hot early versus later in the session, when Dr. Jones shows up"—was transformed into the question: Do Dr. Michaels's hands get hot early versus later in the session *depending on when* Dr. Jones shows up?

This then led to a hypothesis with a prediction: If Dr. Jones shows up early in a session, Dr. Michaels's hands will get hot early in the session; if Dr. Jones shows up later in the session, Dr. Michaels's hands will get hot later in the session. I now thought about how we could test the hypothesis experimentally, the scientific term *operationalize*.

One approach was the possibility that a highly accurate research medium could communicate with Dr. Jones and might be able to work with her over multiple sessions to determine whether she would show up early or late for a given session with Dr. Michaels.

To understand my reasoning, it is helpful for us to recall how after their passing, some deceased people appear to sustain ongoing relationships with their loved ones as well as establish new relationships with people, in both the physical and the spiritual.

For example, in chapter 5 the deceased Susy Smith, besides assisting me in my research, appears to have developed a long-standing and successful relationship with a medium, Joan. We have observed on numerous occasions evidence that Spirit can show up in formal laboratory experiments or people's cars on its own accord. You will recall how Susy Smith apparently showed up in Joan's car, unannounced, and brought along a second, unknown deceased woman, who was later identified.

As it so happened, a medium I will call here Philip had previously done readings with Dr. Michaels and was able to obtain accurate information concerning Dr. Jones. Philip and Dr. Michaels lived in different states, so the readings had been conducted long distance over the phone.

I reasoned: what if prior to a given healing session, Philip could contact Dr. Jones, and together they would decide whether she would show up early or late in the session? Philip would keep a secret log of his communications with Dr. Jones and their decisions about the timing of his visits. Could Dr. Michaels, who would not know when Dr. Jones was scheduled to show up in the next healing session, remember when his hands got hot in the session and record it for subsequent analysis?

If positive results could be obtained in this proof-of-concept private investigation, sophisticated university experiments could be conducted in the future. One of my dream experiments would be to use a state-of-the-art thermographic (infrared) video camera to track the ebb and flow of a healer's hand temperatures during healing sessions. The great advantage of this kind of camera is that it accurately records moment-to-moment changes in temperatures from a distance of at least a few feet; hence, it would not physically interfere with a healer's spontaneous movements or with the temperature sensors attached to the healer's hands.

I further reasoned that (1) if we could perform the pair of events ten times—with Philip and Dr. Jones secretly determining when she would show up in a particular session, and Dr. Michaels then recording when in the given session his hands got hot—and (2) if the results were as replicable as Philip and Dr. Michaels each believed they would be, then ten independent trials would be sufficient to obtain a statistically significant result.

To ensure that the procedures were followed appropriately, I served as the middleman—in other words, I played the role of research assistant. I would contact Philip and ask him to have a secret session with Dr. Jones. I would wait for Philip to call me back to confirm that it had been successful and he had logged when Dr. Jones was to show up. I was kept blind to this information.

I would then contact Dr. Michaels and request that he record when Dr. Jones showed up in his next healing session. Taking into account everyone's busy schedule—and that this private exploratory examination was included as part of Philip's normal clinical work—it took about ten weeks to complete the ten trials.

I did not request that Philip and Dr. Michaels give me copies of their home and office long-distance calling records. I figured that if they were in collusion and wished to fool me, they could easily use other phones, or even email, to keep their cheating secret. However, because I knew both of these individuals reasonably well, I anticipated that their motivation was genuine. Though I would not accept positive results from them as 100 percent legitimate, I had good reason to give them the benefit of the doubt and treat the findings and the experimental design as deserving of further investigation.

The results turned out to be easy to score and analyze. According to Philip, on five of the ten sessions, Dr. Jones said that she would show up early, in the beginning of the respective healing sessions; on the five remaining sessions, Dr. Jones indicated she would show up later. As far as I could tell statistically, the order of early or late for any given session was completely random, like a coin flip.

When I compared Philip's log of the order of Dr. Jones's early and later appearances with Dr. Michaels's log of when his hands got hot in each of the healing sessions, the match between them was 100 percent. The evidence not only supported Dr. Michaels experience, but it did so perfectly.

The observation of a seemingly perfect score should be tempered by the fact that they performed the tests only ten times over a ten-week period. For all we know, Dr. Michaels could have reported different times on the eleventh and on subsequent tests as well. The probability of getting ten out of ten heads by chance alone is small—one out of 1,024, or approximately 0.1 percent. However, it is not tiny, say one out of 1,000,000,024. Therefore, we should not overly interpret the "perfect" score in this experiment.

Does this kind of investigation "prove" that Dr. Jones was actually present? Though this experimental design is clearly novel and highly suggestive, it is not definitive.

Even if we conducted a formal university-based experiment and successfully replicated the basic findings many times using different mediums, healers, and deceased spirits; objective biophysical measures, such as supersensitive thermographic cameras; and convincingly ruling out any reasonable possibility of fraud, the findings would not justify the definitive conclusion that Dr. Jones, or any spirit, deceased or otherwise, had shown up. Other possible psychic hypotheses can be envisioned.

One hypothesis is that Dr. Michaels somehow read Philip's mind, and Michaels's hands warmed up because he psychically expected Dr. Jones to show up in a given session early versus late. We would have to conduct future experiments to determine if healers like Dr. Michaels could read a medium's mind, and if the healer's hands would warm up when he thought the spirits might show up versus when the medium's log said they would.

A related hypothesis is that Dr. Michaels somehow used remote viewing and read what Philip had written in his log. We would have to perform future experiments to determine if healers like Dr. Michaels could read secret handwritten logs. However, future experiments have the potential to be even more innovative and convincing. For example, a medium like Philip could contact a deceased physician like Dr. Jones and instead of deciding when she would show up, they could decide *if* she would show up for a given session.

Science progresses step by step, result by result, investigation by investigation, experiment by experiment. Depending upon specific combinations of positive and negative findings that emerge over experiments, specific hypotheses can be confirmed or disconfirmed, ruled in or ruled out. If I were viewing the Dr. Michaels-Philip exploratory investigation in isolation, I would be inclined to predict that the superpsi hypotheses of possible mind reading or remote viewing were more probable than the presence-of-Spirit hypothesis.

If instead I consider all the information revealed in part II of this book and therefore view the Dr. Michaels-Philip investigation in this larger context, it becomes reasonable to consider the serious possibility that the presence-of-Spirit hypothesis is more probable then either minding reading or remote viewing.

The fact is that if we can see the big picture and hold the whole set of evidence in our minds, we will be able to draw a comprehensive and fair conclusion about the two core questions: are spirits real, and can they play a guiding role in our personal and collective lives?

Meanwhile, once again, I invite you to do a thought experiment.
This time, try to imagine what it might feel like to be in Dr. Michaels's shoes. Try to envision not only what it would be like to experience the apparent presence of your beloved physician grandmother when doing healing work but then to courageously participate in a private exploratory investigation to test the validity of your belief that your grandmother was still with you.

How would you feel waiting to see if the investigation had been a failure or was merely suggestive of your hypothesis, and then you were told that the results could not have been more positive? (It is difficult to score higher than 100 percent.)

And if you are game, you might try to imagine what it was like being in my shoes: (1) Have the superweird personal experience of inviting Spirit to help you for the very first time and feeling your back pain quickly and dramatically be reduced, only to have it return in spades immediately following your asking the question, "Was this a placebo effect?" and hearing "We will take our support away." (2) Later, while conducting a novel private investigation, see highly positive evidence appear that pointed clearly to the conclusion that a deceased physician could show up regularly at healing sessions.

I find it helpful to remember Dr. McCulloch's wise statement that introduces this book: "Don't bite my finger, look where I'm pointing." Though the nature of this proof-of-concept exploration is often confusing and mind-boggling, and occasionally frightening, it is more often than not great fun and even exhilarating.

Meanwhile, I increasingly come in contact with gifted healers who recount amazing stories about how their healings are regularly aided by the assistance of deceased healthcare professionals, spirit guides and angels, and the Sacred itself.

9

A HEALER'S SPIRITUAL LESSON ABOUT THE ILLUSION OF ILLNESS

January 31, 1904
I used to heal with a word; I have seen a man yellow because
of disease, and the next moment I looked at him and his color
was right; was healed. I knew no more how to it was done
than a baby; only it was done every time. I never failed;
almost in one treatment; never more than three.

—Mary Baker Eddy, from *Notes on the Course in Divinity*

If there is one lesson I have learned in life, it is to always be prepared for surprises.

Curiously, despite having taken this simple lesson to heart, I have nonetheless from time to time been unprepared for the degree of novelty and mystery expressed in the seemingly impossible surprises that have jolted my life. Or were they Spirit's designs?

The inspiration for writing *The Sacred Promise* came to me while I was suffering from what might best be described as a case of the superflu. The illness came on quite suddenly, seemingly with no warning, and it quickly incapacitated me. I ended up lying in bed with a fever for six days, followed by another five days of fatigue accompanied by chronic spasmodic coughing.

A week prior to the flu's onset, I had traveled from Tucson to Toronto to give a keynote address on *The Energy Healing Experiments* to an energy psychology conference. While I was there, a series of twenty-four synchronicities involving tigers—yes, the large cats—began, which continued upon my return to Tucson. Let me add here that the phenomenon of synchronicity was, if not discovered, at least scientifically formulated by Dr. Carl Jung, the founder of Jungian psychology and Jungian psychotherapy. A common example of a synchronicity is that you spontaneously think about someone you have not heard from in a while, and then within hours or days, seemingly out of the blue, that person calls you.

Briefly, synchronicities refer to the occurrence of two or more events that happen in close proximity to each other. The co-occurrence of the events is typically highly improbable—meaning that the probability of their occurring, in sequence by chance alone, is very small if not minuscule. The co-occurrence of the events cannot be explained as one event causing another event—Jung called the connection acausal. To explain the nonrandom nature of the co-occurrences, one must hypothesize the existence of some sort of invisible influences or energies, or thread. The co-occurrences often occur at meaningful times in one's life, whereas the actual meaning of the co-occurrences is typically more like dream symbology.

I have been a synchronicity watcher for more than twenty years. I shared my first convincing synchronicity experience, which happened when I was a professor at Yale, in my book *The G.O.D. Experiments*. The chapter was originally intended to be placed within the body of the book, but both my editor and writing partner felt it was too controversial—too "weird"—to be included in the book. (We placed it in the appendices.) However, the fact was that the unique co-occurrences that took place provided compelling evidence for the existence of some sort of Guiding-Organizing-Designing— the G.O.D. the book title referred to—process in the universe as expressed in our personal lives.

By the time I sat down to write the chapter you are reading, I had collected more than one hundred fifty sets of superimprobable co-occurrences of events. I now have come to the conclusion that the tiger synchronicities reported here and the context in which they occurred were showing my wife Rhonda and me that at times, if not in all cases, Spirit can use synchronicities to show the concurrence of its world and ours.

It is not essential for us to review the Toronto Tiger series here. What is important to realize is that as fate would have it, when the flu symptoms began, a set of compelling synchronicities occurred. I note and describe them because they help set the stage for understanding how their convergence was a prelude to the spiritual healings that I experienced. They led me to reexamine my understanding of the synergistic role of Spirit in healings as well as the role that synchronicities serve in revealing the presence of Spirit, acting like a calling card.

There was a set of three spiritual healings that were offered to me, without my awareness, during the course of my illness. The healings were secret in that I was not told that I would be receiving them, and when they occurred I did not know they were being provided.

In each case, the healing treatment was immediately accompanied by a dramatic reduction in my symptoms. Of course, their concurrence with a significant reduction in symptoms might be a coincidence. In other words, the appearance of one, or even two, spiritual healings with symptom reductions might have occurred simply by chance. However, when this happened not once or twice, but three times, the concurrence of a secret healing with symptomatic reduction became sufficiently interesting enough to deserve serious consideration.

This was obviously not a university laboratory experiment. I was sick and I happened to receive a total of three secret healings. Because I practice the philosophy of science in my private life, I was able to witness this demonstration of apparent spirit-assisted healing in the laboratory of my personal life. While such examples may seem subjective and anecdotal, they do fall within the parameters of proof-of-concept observations and experimentations as we explore the paradigm of spirit-assisted healings.

To ensure that we focus our attention on the three healing events or effects—and not the tiger synchronicities (which happen to be quite amusing)—I will begin by describing the first healing treatment that occurred after suffering six days of fever. I will then reveal the unfolding of the synchronicities, and explain how they set the stage for the second and third secret healing events. Since they began when my symptoms were just starting, the discussion will necessarily back-track a bit as we review the first few days of my illness.

Spiritual Healing #1: Secret Treatment and My Fever Breaking

For six consecutive days and nights my temperature fluctuated between 99 and 101 degrees. I took my temperature with a state-of-the-art digital thermometer that displays temperature in tenths of a degree. During this period, I stayed in the bedroom. At night I would close the door. Rhonda graciously and wisely agreed to sleep in a separate room.

It was about midnight on Wednesday, November 12, 2008. At this point I was monitoring my temperature approximately every hour. I began to notice that I was feeling a little better, and at about 1 AM I noted that my temperature had dropped below 98 degrees. I continued to check it every hour, since I was constantly woken up by coughing fits. To my amazement, my temperature ranged between 97 and 98 degrees, never increasing to 99. For some reason, seemingly out of the blue, my temperature had finally broken.

Meanwhile, unbeknownst to me, Rhonda woke up around midnight, hearing a voice in her head saying that she needed to do some healing for me. Rhonda had been raised in a Lutheran and Christian Science home, and was trained in the practice of spiritual healing as a child. Though the scope of her adult knowledge and interests extends substantially beyond conventional Christianity and Christian Science—she has formal training in Reconnective Healing as well—she fervently believes in the fundamental premise that everything ultimately is an expression of Divine Mind, whose existence is perfection, and that the human mind can raise itself and be in harmony with the Universal Mind. This core premise is by no means unique to conventional Christianity and Christian Science; it is fundamental to many energy and spiritual healing traditions around the globe, including mystical Judaism and Sufism.

The reader should understand that I am neither a conventional Christian nor a Christian Scientist and I seriously question a number of their fundamental beliefs and practices. What matters in this chapter is not whether certain Christian or Christian Science beliefs or practices happen to be true but the factual occurrence and timing of the specific healing events and our sincere efforts to make sense of them.

Rhonda told me that she was led to go into her home office and select passages from *Notes on the Course in Divinity*, also called the Blue Book—a collection of teachings of Mary Baker Eddy, founder of Christian Science, as recorded by her students. During the secret treatment, Rhonda noted passages claiming that Mrs. Eddy had experienced miracle-like cures by simply uttering certain phrases to the afflicted. For example, there was the man who had been paralyzed for years but walked after hearing her tell him that "God loves all."

I quote from *Notes on the Course in Divinity*:

March 12, 1907

One of them was one of the worst cripples I ever saw. I was walking along the street—I walked because I hadn't a cent to ride—and I saw this cripple, with one knee drawn up to his chin; his chin resting on his knee. The other limb was drawn the other way, up his back. I came up to him and read on a piece of paper pinned to his shoulder: help this poor cripple. I had no money to give him so I whispered in his ear, "God loves you." And he got up perfectly straight and well. He ran into the house (told name of the people—Allen I think) and asked, "Who is that woman?" pointing to Mrs. Glover (afterward Mrs. Eddy). The lady told him, "It is Mrs. Glover." "No, it isn't, it's an angel," he said. Then he told what had been done for him.

Apparently Rhonda had been hearing the voice for the past couple of nights but had ignored it. For some reason, she decided to listen to it on Wednesday night. She invited her deceased mother, a Christian Science–certified practitioner, along with the late Susy Smith and higher spiritual beings (such as angels), to assist the Divine in this healing service.

After approximately one hour, Rhonda sensed that the treatment was completed, and she went back to the living room and fell asleep. I was startled the next morning to learn of the striking "coincidence" of Rhonda's unplanned and secret spiritual healing and the timing of my dramatic fever reduction.

Was this simply a chance event, or was this connection, as Susy Smith was fond of saying, "too coincidental to be accidental"? I further wondered whether there was a lesson here.

To make sense of what transpired in the two subsequent secret spiritual healings, I will backtrack and examine the onset and course of my fever. (The unfolding of the tiger synchronicities turned out to be important because, as you will see, they eventually related directly to, of all people, Mrs. Eddy and, as incredible as it may seem, her explanation of Christian Science healing.)

The Tiger Synchronicities Pointing to the Big Picture

It was on a Thursday morning, November 6, 2008, that my sporadic bouts of coughing, which had begun a few days earlier, had become so severe that I could no longer speak without triggering a prolonged and uncontrollable fit of coughing. I ended up canceling all of my appointments that day, as I tried to lessen my increasing spasms of coughing with cough drops, fluids, and rest. That evening I noticed that I had a low-grade fever.

At one point I was moved to check the on-demand movie listings. I discovered that the movie *Crouching Tiger, Hidden Dragon* was playing. I had heard about this movie, but it seemed too complex for me to digest in my muddled state, so I simply noted its presence.

I slept quite badly that night. The next morning my fever was more than 100, and I felt miserable. I decided to remain in bed and give my body the additional rest it needed. When I get really sick, the only thing I am comfortable doing is rereading Robert B. Parker's mystery novels, featuring a complex character named Spenser, the basis for the TV show *Spenser for Hire*. He's a private eye in Boston and a former policeman who loves cooking and quotes poetry; his girlfriend is a Harvard-trained psychotherapist, and he is devoted to truth, integrity, ethics, and doing the best he can.

I went into the garage, opened the box that housed my Spenser novels, and pulled out the paperback at the top of the pile. It was titled *A Catskill Eagle*. As I was rereading the novel, I happened to notice that the author used the words *tiger* as well as *crouching* numerous times in the book. Twice the author mentioned an "Asics

131

Tiger gym bag." Once Parker used the word *tiger* in a surprising way: "Each door we opened was crucial. Was there a lady in there? Or a tiger?" And he once referred to someone as "holding the ass end of a tiger."

I began rereading a second Parker novel on Friday (which did not mention a tiger), and continued reading it on Saturday. I also spent part of Saturday watching a college football game between the University of Alabama and Louisiana State University at LSU, which I discovered happened to be the home stadium of the Tigers. There was even a large portrayal of a tiger at the LSU fifty-yard line.

Meanwhile, the other college football game on another channel involved Penn State University and Iowa State University at ISU. To my surprise, I heard the announcers having a conversation about how every five minutes or so someone was bringing up Tony the Tiger. Two tiger mentions on television at the same time connected to football? I began to take notice.

I scribbled some notes about these potential synchronicities on the front page of *A Catskill Eagle*. I knew that I would forget the details and the timing if I did not record the events as they were happening, especially since I was groggy and not feeling well.

I then discovered that a James Bond movie was playing on another channel. When I switched to it, the scene in progress involved Bond being picked up in a car driven by a female agent with the bad guy's car chasing them. The scene took place in Japan. Bond spoke by phone with a foreign male operative. To my amazement his code name was, you guessed it, Tiger!

Think about this: there were now three tiger mentions on television in the same time period. I made a point of adding the James Bond reference to the growing list of synchronicities.

The next morning, Sunday, November 9, 2008, I began rereading the third Parker novel, titled *Taming a Seahorse*. I noticed that one of the lead characters worked at a place called Tiger Lilies. The establishment answered the phone "Tiger Lilies" on page twenty-three and by page eighty-seven, there had been six tiger mentions. As I would soon learn, not only was the word significant here but its specific connection with lilies was important as well.

I took a break from reading and turned on the TV. I noticed that on a Spanish-speaking station, the program that happened to be playing was *Fútbal Mexicano Tigres*. A soccer team called the Tigers was playing? Hmmm . . .

Then Rhonda came into the bedroom and told me that a documentary about tigers was on TV called *Growing Up Tiger*. I pondered the improbability of a tiger documentary being on at that moment. Though I was having a hard time concentrating—between the aches and pains, dizziness, nausea, and general malaise—I could not resist watching the program. It featured two baby tiger cubs during their first year of life. The male tiger was named Sergeant; the female tiger was named Tiger Lily. That's right, Tiger Lily.

First I read about Tiger Lily in the Parker novel, and then I saw a tiger cub named Tiger Lily the same day. What's the probability? The answer is: miniscule. I could not wait to tell Rhonda about the apparent Tiger Lily connection.

After the documentary ended, Rhonda came back to the bedroom all excited. Before I could say anything, she reminded me that her beloved dog, a beautiful Borzoi who had died over a decade ago, was named Lila. Rhonda and I had been thinking about adopting one, and she had suggested that we name our Borzoi Tiger Lily—Lily, for short. Rhonda's spontaneous inspiration about our future dog's name could not have been more propitious.

As it so happened, one more remarkable Tiger Lily synchronicity occurred the next morning. I received a surprise phone call from Robert Stek, PhD, who at the time was in Regina, Canada, giving a presentation on our paper about Sir Arthur Conan Doyle, his history involving spiritualism, and our latest personal afterlife research involving Sir Arthur. I wrote about a string of tiger syn-chronicities in the second of two books, still in progress, and mentioned a few of them involving Bob. The two books illustrate how self-science can be used to discover and document the existence of strings of synchronicities in our personal lives—what I call super-synchronicities—and how science and spirituality are mutually advanced by examining them.

Bob had called not only to share what had happened at the Inter-national Sherlock Holmes Conference but also to tell me about a curious synchronicity involving the flag of Saskatchewan, Canada, of which Regina is the capital. In the center is a specific flower, which just so happens to be a tiger lily.

I was beginning to sense that there were too many tiger syn-chronicities occurring during my illness not to have some sort of meaning, but I had no idea what that was. At that point I had had the fever for four days, longer than I had experienced in years.

By Wednesday, the sixth day of the fever, I was fit to be tied. I had learned various energy self-healing techniques over the years, which I could have applied if I had a localized physical symptom, such as back pain or bleeding. Instead I found that among the fever, dizziness, nausea, pain, and fatigue, I could not focus my attention and use any of them. All I could do was watch a little television and reread Parker mysteries. Rhonda had our car repaired on Wednesday, and noticed in the window of a CVS drugstore a cute stuffed tiger. She bought it for me, and I named it Tiger Lily.

And then on Wednesday night around midnight, my temperature suddenly broke, and the next morning, as previously discussed, I dis-covered that I had received an unexpected spiritual healing from Rhonda. Notice that this spiritual healing had no obvious connection to tigers. Moreover, it never crossed my mind or Rhonda's that some-how the unfolding tiger synchronicities might be connected to a spiritual healing.

Spiritual Healing #2: Secret Treatment and My Night-Coughing Breaking

As far as I could tell, with the breaking of my fever, the tiger synchronicities stopped, too. I ended up rereading a total of fifteen Parker novels during the course of my superflu, and only the first and third mentioned tigers.

From Thursday through Saturday, I felt like I had been run over by a fleet of trucks. Though I was fatigued, the absence of the fever was a blessing. However, my coughing got worse, especially at night. I would wake up every fifteen to forty-five minutes with bouts of uncontrollable coughing. Rhonda continued to sleep in another room. The hacking would still sometimes wake her as well.

On Saturday night sometime around 1 AM, I had a particularly severe bout of coughing. It could well have been the worst I had experienced. After that I noticed that the urge to cough had greatly subsided. In fact, after an hour, I found I could finally go to sleep, and I actually slept until around 8 AM with virtually no coughing. I could not wait to tell Rhonda.

When I joined Rhonda in the living room, I learned that the night before she could not wait to talk with me. She said that she again had heard a voice saying that she needed to do a treatment. She went into her home office and this time selected, seemingly randomly, a section from Mary Baker Eddy's classic *Science and Health with Key to the Scriptures*.

Rhonda could not believe what she was reading. Mrs. Eddy was using a metaphor of a tiger!

At first I thought Rhonda was either joking or delusional; I quickly learned my initial thoughts were completely in error. I quote the complete paragraph with the tiger metaphor below:

Without the so-called human mind, there can be no inflammatory nor torpid action of the system. Remove the error,

and you destroy its effects. *By looking a tiger fearlessly in the eye, Sir Charles Napier sent it cowering back into the jungle.* An animal may infuriate another by looking it in the eye, and both will fight for nothing. A man's gaze, fastened fearlessly on a ferocious beast, often causes the beast to retreat in terror. This latter occurrence represents the power of Truth over error—the might of intelligence exercised over mortal beliefs to destroy them; whereas hypnotism and hygienic drilling and drugging, adopted to cure matter, is represented by two material erroneous bases.
(*italics added*)

At that moment, Rhonda was in awe. Her husband was suffering with a bad case of the flu. She had waited six days to perform a spiritual treatment—you will recall that I had not asked her for healing treatment. She ended up doing the first treatment (and two subsequent ones) without my awareness. At the time she was performing the first treatment, she had no direct feedback that this was occurring at the same time that my fever was breaking.

My impression has been that Rhonda is a reluctant healer, respectful of other people's beliefs and values. According to Rhonda she felt the need to provide me with a treatment only because she heard a voice in her head—this was her personal experience. And to her surprise (and mine), her efforts were validated with the breaking of my fever.

Now, try to imagine what it was like for Rhonda to awaken four nights later with the feeling that she must do a second secret treatment. She was hearing her husband coughing like mad. She went into her home office, thumbed through *Science and Health*, and began reading a section. Two pages later, what did she read but a specific passage referring to tigers and the process of healing!

This time Rhonda could tell that the timing of her treatment was associated with the reduction of my symptoms. She could hear that I had stopped coughing. Both Rhonda and I slept the rest of the night. It was only upon my waking in the morning that I learned of Rhonda's second treatment and her reading about the tiger.

Later that day I went back to Rhonda's office and reread the passage. The tiger synchronicities were pointing to the message of this passage for me, about the human mind and its beliefs that are in error and staring down this wild beast: "This latter occurrence represents the power of Truth over error."

As I said, I had been contemplating writing a book about the subject of our partnership with Spirit but had been held up by the lack of hard-science experiments to verify what I knew in my heart and was experiencing in my life: that Spirit is there waiting to help us. Did I dare to go ahead using self-science as one of the measuring sticks? But I was also amazed by the delivery system. I might have thrown the *I Ching* and gotten a similar message, but apparently Spirit chose to show me how our lives are the ultimate metaphor for Spirit and its effects.

The Tombstone Tigers

I felt the need to celebrate the apparent emergence of some sort of lesson involving not only spiritual healing but also a better understanding of Spirit's partnership with us, a gift that Rhonda and I were receiving individually as well as collectively. Moreover, our wedding anniversary was in three days. I suggested to Rhonda that, if she was willing to drive, we go to Tombstone, Arizona, have lunch at a low-key restaurant, and maybe purchase something to honor our upcoming anniversary coupled with the still-mysterious healing/tiger lesson that seemed to be unfolding.

Rhonda drove us to Tombstone. We had lunch at the Pioneer Grill, known locally for its wonderful home cooking, and then walked over to Tombstone's best Native American art store, Arlene's. We had previously purchased a number of pieces from there.

Native Americans do not, as a rule, carve or paint tigers, since tigers are not indigenous to North America. However, Rhonda asked the shop clerk what probably seemed to him to be an utterly foolish question: did they have any pieces of art with tigers?

To our amazement, the clerk explained that just last week they had received a set of kachina-like sculptures and one of them was of a tiger! In fact, the owner had requested a kachina tiger sculpture—something he had never done before.

This was ridiculous. A tiger sculpture in a Native American store? Unfortunately, the piece was too costly, so we did not purchase it, but they did allow us to take some photos.

The owner of Arlene's has a second store that contains contemporary art. We visited that store and discovered a small, carved tiger statue that I purchased as a memento. I named the statue Mary, in honor of Mrs. Eddy, but also for Rhonda and her connection.

When we returned from Tombstone, I was wiped out. I had had more than enough adventure for one day. I turned on the television and noticed that the movie *Evan Almighty* was in progress. *Evan Almighty* is a spiritual comedy about God, played by Morgan Freeman, who asks Evan Baxter, played by Steve Carrell, to build an ark.

I had watched *Evan Almighty* at least ten times, partly because John Debney—who wrote the music for it, as well as for *Bruce Almighty* and *Dragonfly*—is a dear friend, and Rhonda and I had the privilege of being present in Studio City when John conducted the orchestra and chorus that put the music to the movie.

The scene I chanced upon had tigers in it. I had not paid attention to the tigers in this movie before. I later discovered there were at least five scenes that included tigers. What was meaningful for me was the timing of seeing *Evan Almighty* with new eyes shortly after having placed Mary the tiger on my desk.

Just as the movie was ending, Rhonda came into my office and told me that the segment in the documentary *The Brain*, featuring

Dean Radin, PhD, and me, was about to air. I switched to the History Channel. I had seen the film once before, but it had been a while. In fact, I didn't remember what Dean had talked about concerning precognition research or what I had said about afterlife research.

As we watched the segment, to my delight I witnessed Dean discussing the relationship between psychic abilities and golf, and he used Tiger Woods as the exemplar! They even included a scene with Tiger making a putt.

I wondered, why the reappearance of the tiger synchronicities? I seemed to have had my ah-ha moment. Was there perhaps more for Rhonda and me to learn?

It was clear that I had experienced on two occasions—once with the fever and once with a coughing fit—an uncanny timing between the dramatic remissions of my symptoms and secret spiritual healing treatments by Rhonda. The combination could have been a double coincidence, two chance timings of spiritual treatments with symptom remissions. But a triple concurrence statistically would suggest that it was really too coincidental to be accidental.

As it turned out, there would be a third treatment. And it would be the charm, occurring on the morning of our wedding anniversary.

Spiritual Healing #3: Secret Treatment and My Night-Coughing Ending

The adventure on Sunday was too much for my weakened immune system, because that night my coughing reappeared in full force. Rhonda heard me coughing but apparently was not moved—or was not prompted—to do another treatment. True to form, I did not ask her for a treatment. My coughing continued on and off during Monday and that night, they were epic. And Monday night Rhonda was inspired to do a third treatment. Once again, in the middle of the night, my coughing suddenly subsided.

On Tuesday morning, our anniversary, Rhonda confessed that she had done a third treatment. At this point, I felt the need and responsibility to learn exactly what was she doing. I had never asked her to write up, in detail, what her personal healing treatment experiences entailed.

Below, in Rhonda's own words, is what she experienced. For the sake of clarity, I've inserted some commentaries where appropriate.

November 17, 2008
I went to sleep on the couch again tonight, wanting to give Gary space to relax and sleep peacefully.

At some point, I was awakened by him coughing violently. I immediately and with authority declared out loud but in a whisper, "No, no, no! That is not the truth about you. Gary is not sick; he has never had a cold or a cough. And I am not impressed by or afraid of what is being presented here. It is a lie, and I absolutely do not accept or believe what I am seemingly hearing."

For those of you who are not familiar with Christian Science philosophy, practitioners believe that the essence of a person is spiritual, not physical, and that there is no disease in their spiritual essence. What is seen as physical disease is interpreted to be an illusion in the sense that what appear to be symptoms do not represent the true essence or reality of a person.

Of course the symptoms are occurring: the term *illusion* refers to our interpretation of the meaning of the physical symptoms, not the existence of physical symptoms per se.

Instead, the presence of physical symptoms is viewed as reflecting the consequences of errors in our understanding the essence of reality. Practitioners believe that when they forcefully and honestly connect with the higher essence of reality—what they call "Divine Truth"—

that the true nature of a person can be expressed. In this sense they are "staring down the tiger of disease."

I then mentally, completely turned my thought away from Gary, the sense evidence [the evidence we experience with our senses], and anything other than God. I let the all-ness of God completely [Rhonda's underlining] fill my consciousness, leaving no room for anything else and continued with a metaphysical treatment of prayer that went something like this:

God, Spirit is All (All-in-all)—there is nothing else. God fills all space, and he created all that is, and his creation is spiritual and perfect, changeless throughout all eternity. Gary is a spiritual idea held by God in Mind—perfect in form, function, and outline. There is no inaction, over-action, or reaction in the Divine Action. And in the equipollence of God all pressures and temperatures are even because they are governed and controlled by God.

Notice that this prayer is very different from the typical practice of energy healing as well as the various prayer-based spiritual healing practices. At no time was Rhonda consciously sending me loving healing energy. At no time was she consciously attempting to use her intentions and energies to selectively improve my immune function or suppress my coughing, or beseeching God to fix me, or asking God to return me to health. Also, Rhonda was not employing Reconnective Healing techniques at that moment.

Instead, Rhonda was making a series of philosophical statements about her belief in the ultimate essence and reality of the Universe, or the unity and oneness of all that is—and she was filling her consciousness with this core philosophy. The next part was news to me:

This would normally have been the end of my treatment, but since my mother, who passed a few years ago, had been a full-time practitioner/healer in Christian Science, I asked her to assist as well. She had shown me many times that she was still very much alive and involved in my life.

To honor Gary and his scientific interest in not only those passed on the other side but also the possibility of angels and guides as well, I invited Susy Smith and any councils, angels, guides, or beings of any kind, if they heard my request, to please help Gary be free and sleep comfortably that night.

For the record, after this secret treatment, I experienced the most peaceful and restful sleep. Moreover, my sleep was virtually perfect the subsequent nights. My coughing fits, for all practical purposes, ceased following Rhonda's third and final treatment.

In my six decades of life, I had never experienced or witnessed anything quite like this healing sequence. In the process of coping with this eleven-day flu, I witnessed not once, not twice, but three times my severe symptoms subsiding immediately following three unrequested, secret spiritual healing treatments.

In a deep sense, I was blind to the actual occurrence of the treatments. The healings occurred without my conscious awareness. Once again my personal life was arranging itself to function as a living laboratory for proof-of-concept discovery and learning, for myself and, in this case, for Rhonda—and now, potentially, for you.

But What Does This Mean?

My personal healing experience was obviously not a controlled laboratory experiment conducted under the auspices of a university. It was, however, a true-life experiment, the kind that really matters. Scientists can design elegant laboratory experiments that do not generalize or apply to the real world. The real test is whether the theorems discovered in formal laboratory science can explain the operations of nature and the real world, or can the real world give inspiration to new theorems, like in Einstein's case, when the relative movement of a train and people walking on a platform contributed to his theory of relativity.

The heart of science is observation of the natural world, and paradigm shifts come about when previous explanations or theories do not adequately account for a whole set of observed phenomena. Since it involves observation and scientific analysis, using myself as the subject in these self-science explorations of Spirit and its possible healing effects is the first step, with a hypothesis being the second, and then laboratory-based replication the third.

I could spend pages detailing the various possible explanations of the three healings that occurred, including:

1. Spontaneous remissions accompanied by highly improbable but nonetheless chance coincidences

2. Fraud and/or misperception by Rhonda and/or me

3. Subtle cuing on Rhonda's part that somehow might have been picked up by me, a so-called subtle placebo effect

4. The direct effects of Rhonda's focused love and compassion on my biochemistry and physiology—in other words, the effects of her loving energy and intentions on me

5. The assistance of Spirit—including a deceased practitioner/ healer, angels, and the Sacred itself

Part of the reason I seriously entertain some version of the fifth listed explanation is because of the presence of the uncanny tiger synchronicities surrounding the healings and how they persisted even after I seemed to have gotten the message. It seems this happened so that I could be exposed to a philosophy of healing and of the convergence of the Spirit world and ours as the basis of this Sacred Partnership.

The persistent occurrence of the tiger synchronicities, interwoven with Mrs. Eddy's writings, speaks to the strong possibility that something more complex, intelligent, and awe-inspiring was taking place. The truth is, if Spirit can be involved in physical healing, we know next to nothing about how it works scientifically.

Using Mrs. Eddy's label, which introduces this chapter—and this is especially the case scientifically—we are like babies in terms of our knowledge and understanding of the role of Spirit in health and healing. However, we do not have to remain in our infancy. We have the choice, and the opportunity, to use the gift of our discovering minds to address these most precious and profound questions.

If we are brave enough to ask these questions, we will stand up to the tiger and look it in the eye.

10

THE ROLE OF SPIRIT IN EMOTIONAL HEALING

It is belief that gets us there.

—from the movie *Dragonfly*

Whether explicitly or implicitly, the motivating force behind all the accounts of apparent, spontaneous Spirit assistance shared in this book has been love.

Love is the common denominator that has propelled spirits (1) to reveal their presence, (2) to provide useful information, if not protection, and sometimes intentionally to hide information (part II), and (3) to participate in healings and our understanding of them (part III).

The same conclusion—that love is the driving force—seems to apply to the predominant appearance of synchronicities, especially when they appear to have been mediated by Spirit. Again, synchronicities are the highly improbable occurrence of two or more events in close proximity, where one is not causing the other but suggests the existence of some invisible influences or energies causing their co-occurrence.

What you are about to read is a true account of a series of synchronicities involving dragonflies that appear to have been spirit mediated and ultimately to involve a spontaneous case of spirit-mediated healing of a deep emotional trauma.

How would you feel if you were a surgeon's assistant and your wife, who had a serious disease, attempted suicide by cutting her throat, and you found her dying, attempted to save her life, but failed? What would go through your mind after living with the guilt of your failure if one day your deceased wife appeared in an utterly convincing way and told you that it was not your fault that she died?

When I first saw the movie *Dragonfly*, I had no idea that it would lead to a series of synchronicities that directly related to spiritual guidance and emotional healing. Though the movie was profoundly moving and meaningful to me, I had no inkling that its unforgettable message would soon blossom into a flower of increasing and unimaginable synchronicities.

The plot of the movie involves two physicians who are very much in love: the husband, Dr. Joe Darrow, who specializes in emergency medicine, and his wife, Dr. Emily Darrow, who works in pediatric oncology. Emily is pregnant, and they are eagerly awaiting the birth of a baby girl. Emily goes on a Red Cross mission to South America to help young children. A horrific accident occurs, and everyone including Emily are presumed dead.

After the accident, Joe begins experiencing strange happenings at home involving one of his wife's favorite creatures, the dragonfly. Portrayed as a skeptic, Joe finds these dragonfly-related events, suggesting his wife's continued presence in his life, impossible to believe.

THE SACRED PROMISE

While fulfilling his promise to Emily to take care of her surviving pediatric patients if anything ever happened to her, Joe finds that various children whose hearts had stopped beating and had to be resuscitated are having unusual near-death experiences (NDEs) that seemingly involve his deceased wife. The children claim that Emily is calling out to Joe from the other side with a message; however, they do not know its precise meaning.

I remember closely listening to the hauntingly beautiful music accompanying these scenes and wondering about the composer. Though the story was sensational at times, the movie nonetheless accurately portrayed the controversial claims that people who experience NDEs can sometimes receive important messages, purportedly from deceased people—whether they are their beloveds or total strangers.

As Joe's world is turning upside down and spinning out of control, he learns that a nun, Sister Madeline, has been collecting data on the children's NDEs. Joe eventually meets Sister Madeline in a chapel lit with hundreds of candles. Sister Madeline explains scientifically and poetically how these afterlife experiences are often real.

Toward the end of this spellbinding scene, she makes a simple statement—a take-home message—that touches Joe deeply. Sister Madeline says, "It is belief that gets us there." This line was both a reminder and revelation to me. It recalled Yogi Berra's catchy phrase, "If I hadn't believed it, I wouldn't have seen it."

What the movie expressed was the general principle that, although knowledge and skills are essential for any endeavor, it is the energy of personal belief that motivates and inspires us and, you could say, attracts the conditions that help fulfill our tasks, whether mundane or otherworldly. Simply stated, *though knowledge can take us there, it is belief that gets us there.*

Dragonfly became one of my favorite spiritual-oriented movies. However, I had no idea that this movie would precipitate a

147

completely unanticipated set of future events, the capstone being my seemingly chance meeting with a giant dog, his rescuer, and her mother, whose friend had a remarkable NDE experience that transformed her life—and ultimately, mine.

The Late Susy Smith and Her Messages from the Other Side

This story is about to get a bit complicated, so please bear with me. Except for the changing of certain names and incidentals to preserve anonymity as needed, this is a completely factual account.

It was early morning, around five, and I was preparing to leave for a one-week lecture tour in California. My scheduled talks included a presentation at the San Francisco chapter of the Institute of Noetic Sciences, where I was speaking on *The Afterlife Experiments* book and giving a special address at a Deepak Chopra–sponsored weekend event at La Costa Resort and Spa.

Because I was to be out of town for the week, I had various personal and professional details that needed attending. Before leaving for the university and then the airport, I checked for any emergency emails and discovered an unexpected one from Joan, the medium from chapter 5. You will recall that Joan had originally contacted me by email claiming that she was receiving afterlife communications from Susy Smith, and she subsequently helped discover Susy's double-deceased paradigm. Joan and I had since stopped our private investigations. However, every now and again, Joan would spontaneously email me with messages allegedly from Susy.

In this particular email, Joan mentioned that Susy was bringing up the movie *Dragonfly*, and that this was important. Also, Susy was claiming that I would shortly be meeting a well-known person who was important to my work. I remembered that I was scheduled to meet someone who said he wrote music for the movies. His name was not familiar to me and I had not yet looked him up on the internet.

He had read *The Afterlife Experiments* and was sufficiently intrigued that both he and his wife requested private readings with one of the mediums I tested and reported on in the book. (I have been asked not to disclose the name of this particular medium. It was not Joan, nor had she yet met this medium.) He lived in California, and we decided to meet during my lecture tour.

As I pondered Joan's email, the thought popped into my head, could this person somehow be related to the movie *Dragonfly*? I quickly Googled his name—John Debney—and quickly found that he was the composer of the hauntingly beautiful soundtrack! I cannot adequately express my excitement in words. I had just discovered that I was shortly to meet the composer for not just some movie I liked but a movie whose music spoke to my soul.

But more important, I had been led to this discovery via an email from a medium who lived more than a thousand miles from me and who claimed to have received this information from the late Susy Smith! Was Susy up to something new?

In a moment of enthusiastic gratitude, I emailed both Joan and John and told them how I was led to discover our *Dragonfly* connection. Joan wrote back and reminded me that she too had a connection with the movie *Dragonfly*. Joan had discovered that she spoke a language she did not understand and was never taught. It took her years to determine whether it was gibberish or a real language. A scholar of indigenous languages in South America discerned that the language was a relatively unknown dialect called Yanomami. The language spoken by the tribe that played a critical and dramatic role in *Dragonfly* was—you guessed it—Yanomami!

I now had three *Dragonfly* synchronicities: the purported communication from Susy, my upcoming meeting with the composer of the score for *Dragonfly*, and Joan's extraordinary language connection to the movie. There was definitely something brewing. My mind was soaring with possibilities. Could it be that my dear Susy was preparing me to meet the composer? Or could it be that the theme of the movie—getting a message from beyond—was somehow related to my research on life after death?

I had no idea. I did not have time to ponder such possibilities further as I prepared for my trip. I went outside to the mailbox. Among the letters and catalogues was a gold-colored, bubble-wrapped envelope. To my amazement, it was from John Debney! Had John sent me CDs of his music? Could one be from *Dragonfly*? My hands trembled as I opened the envelope. Sure enough, inside was a collection of CDs, and one of them was the soundtrack.

Holding back tears, I opened the CD and placed it in my home stereo system. Out poured the glorious music John had composed for *Dragonfly*. I raced back to my computer and sent a follow-up email to John and Joan. I shared all of these *Dragonfly* synchronicities with them.

At this point, I knew that something special was happening. I wondered what could possibly happen next.

A week later, the seemingly impossible happened in spades.

The Rescued Great Dane and a Spirit-Mediated Healing

Though my trip to California was memorable, it merely set the stage for what would transpire when I returned to Tucson. It turned out that I did not actually meet the Debneys on this trip. They had to cancel due to an unanticipated illness, and we rescheduled for the following month. This turned out to be propitious. By the time we finally met, I had inadvertently been led to firm evidence indicating that the foundation of the movie *Dragonfly*—that messages can be spontaneously conveyed during NDEs—was real.

Here's what happened.

I returned to Tucson on a Monday afternoon. I had a dinner meeting scheduled at a Chinese restaurant not far from my house with one of my university colleagues. It is worth noting that we had at least fifty dinners out together over a five-year period, but only that fateful night did we eat at that Chinese restaurant, or any Chinese restaurant for that manner. (In the process of editing this chapter, I was inspired to add the name of the restaurant. The Golden Dragon. It was only upon typing the name that I saw an apparent dragon/dragonfly connection that I had previously overlooked.)

Because I happened to arrive at the restaurant fifteen minutes early, I had the unanticipated opportunity to meet the biggest dog I have ever seen. And curiously enough, connected with this dog was a real-life NDE story even more unbelievable than the fictional story in *Dragonfly*.

Around the corner from The Golden Dragon is a Starbucks with outdoor tables. Three women sat with a gigantic Great Dane—which are typically large, but this one was huge. Being a dog lover, I felt the need to meet him. As I quietly and gently approached the dog, he showed an obvious desire to make contact with me (which included licking my face—quite an experience given the size of his tongue).

The young woman holding the dog's leash was surprised and pleased. I introduced myself and asked about her dog. She explained that she had recently adopted him from a Great Dane rescue center in Phoenix and that he apparently had been abused and was shy around most men. For some reason the dog sensed I was safe and responded with great affection.

I explained that I taught at the University of Arizona. The young woman said, "Really! I am a junior there."

Then one of the older women asked, "Are you the Gary Schwartz who wrote a book called *The Afterlife Experiments?*"

I said, "Yes, why do you ask?"

She replied, "Because I was recently given your book, and I'm reading it right now."

I asked who gave it to her, and she said, "Jerry Cohen, the CEO of Canyon Ranch. Do you know him?"

"Yes. In fact, one of the experiments in the book is called The Canyon Ranch Experiment because it took place there. How do you know Jerry?"

"I've worked at Canyon Ranch for years," she explained. She then said, "You know, a friend of mine had an experience that is more extraordinary than any account you provide in your book."

Did I hear her correctly? Was her friend's experience more extraordinary than *any* account in my book? Needless to say, I was intrigued.

I said, "How interesting! Would your friend be willing to speak with me?"

She replied, "Actually, my friend wrote up an account of her experience over six months ago and has wanted to contact you, but she's been afraid."

"Really?" I asked. "Let me give you my cell phone number. Please tell your friend I would be happy to meet with her and discuss her experience."

The next day I received a phone call from the woman's friend. Her story turned out to be so unbelievable and evidential that I requested that she and her husband meet with me to record the events for future research. What made her story so unbelievable was that it related directly to the movie *Dragonfly* and the possibility of NDE contact with the deceased.

A few days later we held the meeting and I videotaped it. What I can share here is a synopsis of the key points of her story, changing names plus a few details to preserve the anonymity of the people involved. We will refer to her as Lynn. She was suffering from a repetitive-motion injury and was about to have it surgically repaired in Phoenix. During the anesthesia process, Lynn apparently had a severe allergic reaction and she had to be resuscitated. When she stabilized, the surgical team decided it was safe to perform the operation.

In the recovery room, the lead surgeon informed Lynn and her husband that, while there had been a problem during the administration of anesthesia, it was caught in time. He also said that something unusual and upsetting had happened immediately following the surgery concerning one of the surgical team members, but that the operation itself was successful and uneventful and he expected Lynn to have a smooth and full recovery. The surgeon did not offer any details about what transpired immediately following the surgery, and Lynn and her husband did not request additional information at the time.

Meanwhile, Lynn began to vaguely remember that during the surgery she had had a surprising and disconcerting out-of-body experience where she felt that she was floating above the operating table. She also remembered being awake at one point and saying something to one of the surgeons. She had relatively little memory of what else transpired. Lynn was embarrassed to talk about this strange experience, and though she did tell her husband about it later, she did not share it with her surgeon.

Here's where the story gets eerie.

A few weeks later, Lynn and her husband happened to see the movie *Dragonfly*. As she watched the various scenes with children having near-death experiences and supposedly receiving communication from the other side, she began to wonder if her own out-of-body experience was somehow accompanied by afterlife events that might have upset one of the members of the surgical team.

After mustering her courage, Lynn called the lead surgeon and asked him if he would tell her what had happened following the surgery. What he told Lynn and her husband would have been completely unbelievable had she not been prepared by seeing *Dragonfly*.

Apparently she was lying down on the operating table, still under the anesthesia, when she suddenly bolted straight up and looked around room, scanning the five people in surgical masks. She then selected one person in particular and said: "There is a woman here. Her name is Sarah. She has a message for you. She wants you to know that it is not your fault that she died."

Lynn then fell back on the table, apparently unconsciousness again. The person to whom she directed the message fled the room in tears.

Lynn later learned that the person who fled the room had indeed been married to a woman named Sarah (name changed to preserve her anonymity). According to the surgeon, Sarah had been suffering from an incurable and deadly disease, and she had become severely depressed. Sarah decided to end her life by slitting her throat. Sarah's husband happened to come home as she was bleeding to death. He attempted CPR but was unsuccessful. As a result, he lived with extreme guilt, believing that had he come home sooner, he might have been able to save her life.

Just imagine the string of synchronicities that drew all three of us together for our own personal epiphanies: the surgical team member who witnessed his wife's suicide and then had a heavily sedated patient relay a message from her that it wasn't his fault; the patient who had the out-of-body experience and whose memory of it was jogged by seeing a movie about messages from the beyond; and the scientist whose seemingly chance encounters led him to this woman and her story, the one person she had meant to tell about it. The combination of these three sets of synchronicities is breathtaking to say the least. It would appear that a highly improbable, if seemingly impossible, set of events and connections was unfolding among a diverse and widespread group of people over extended periods of time.

Some Phenomena Can Only Be Observed in Real Life

Whereas some kinds of research on spirit-assistance can be brought into the laboratory, other kinds can only be observed when they spontaneously occur. In the same way that we cannot bring shooting stars into the laboratory—we can only observe the tracks of their light when they occur—some synchronicities, like those above, only happen in the outside world. We must be fortunate enough to observe them when they happen and chronicle them carefully. We have no control over these observations; we must record and celebrate when they occur.

The *Dragonfly* synchronicities provide an example of self-science par excellence, where the observations are compelling to our minds while they touch our hearts. Certain anecdotes are worth their weight in gold. I later learned that the surgical assistant who received the unexpected message of solace and love from his wife via his patient was greatly healed by the content and context of the gift.

Lynn had a profound spiritual experience that she cherishes to this day.

For me, the message was more complicated.

If I had met Lynn as described above and heard her story, I would have been amazed without all of synchronicities that led up to it.

So, what was the additional lesson here?

I realized that I was being shown how Spirit actually operates in our lives, and that synchronicities are often its calling card.

But the carry-home message for me was that self-science is the next frontier for verifying the existence of Spirit and its operation in our lives and for pushing forward into this uncharted territory. Sometimes we must revisit the experience and see it anew to discover even greater connections—the *Dragonfly*/Golden Dragon connection being a case in point.

Yes, the knowledge of science provides me with techniques and skills. But I believe that the power of science, when creatively and wisely applied, can help us better understand who we are, why we are here, and what our potential is, and that motivates me to do this work.

I personally love science because of its extraordinary power to serve humanity and the planet and to serve and influence each of us as we live our daily lives. Under the right circumstances, science is one of humanity's greatest friends.

My belief in science is an evidence-based belief. I have witnessed the power of science time and time again. Yes, the scientific method can be, and has been, abused, and yes, there are risks involved in applying it. However, there is no substitute for it.

Fortunately, although we can't bring shooting stars into the laboratory, we can sometimes bring spirits into the laboratory. And under the right conditions, they can reveal themselves in breathtaking ways, even using state-of-the-art electronic technology (revealed in part IV).

I sometimes think of love in scientific terms:

Listening (with a questioning yet open mind)
Operationalizing (translating observations into hypotheses that can be tested)
Verifying (replicating and extending observations under ever more controlled conditions)
Explaining (formulating interpretations based on replicable observations)

You can learn to L.O.V.E. science, too, and bring it into your everyday life. If nothing else, *The Sacred Promise* affords us the opportunity to see how science can lovingly bring Spirit and synchronicities into our everyday lives.

Meanwhile, we can only wonder what Susy—who apparently sometimes acts as my adviser and protector on the other side—may have experienced while preparing me for these *Dragonfly* adventures, or what Sarah may have felt—possibly brought there by Susy—being able to give her husband the solace and love he needed.

PART IV

The Promise of a Greater Spiritual Connection

11

THE CASE FOR SPIRIT GUIDES

The reason angels can fly is because they take themselves lightly.

—G. K. Chesterton

Until quite recently, when I heard the words *spirit guides* (including the subset *angels*), I thought about Santa Claus and smiled. For me, spirit guides were like Santa Claus or other lighthearted mythical beings, such as the Easter Bunny. Most mature adults know that Santa Claus and the Easter Bunny do not exist, and they apply the same reasoning to spirit guides and angels. This is what I was taught even as a child.

Moreover, the vast majority of scientists today hold the firm conviction that spirit guides and angels are silly fictions created by misguided minds. Scientists, as a general rule, are convinced that if an adult person believes in spirit guides or angels, he or she must be ignorant, illogical, foolish, or deluded, if not crazy. I suspect many of you would be so convinced, too.

There was a store in Kirkland, near Seattle, Washington, called Reasons to Believe, which sold handcrafted Santas, Santa carvings, Russian Santas, clay-sculpted Santas, Santa figurines, and Santa ornaments. The name of the store did pique my interest. After exploring the large inventory of artistic and mythical expressions of the Santas displayed in the store, I asked the owner point-blank if he believed in Santa Claus.

He said, "No, I don't believe in Santa; I believe in the spirit of Santa."

When I asked the gentleman to explain himself, he said that he believed in the idea of universal and responsible giving, reinforcing those principles, especially in children, and encouraging people to live their lives for the greatest and highest good, and then helping them make their dreams come true.

What an interesting and inspiring belief.

I realized that if someone had asked me back then, "Do you believe in spirit guides?" I could draw on the store owner's inspiration and say, "No, I don't, but I believe in the spirit of spirit guides." Similarly, I would've said, "No, I don't believe in angels; I believe in the spirit of angels."

What I would mean by these statements is that I believe in the philosophy of universal caring and giving, of nurturing and protecting others, especially loved ones. And in regard to Santa, I further believe in the value of sometimes giving anonymously, and when possible giving in such a way that the recipients are not aware that they are receiving a gift. An example may help.

A former undergraduate student of mine, whom I will call Patricia, once confessed that during Christmas, when parking lots were packed, if a car filled with children was following her and a space opened up, she would sometimes intentionally pass by it, giving the space to the family behind her. The family never knew that Patricia was serving, in this instance, as an anonymous giver, an invisible angel, so to speak.

I have been forced to reexamine my past conviction that spirit guides and angels are like Santa Claus. In fact, as you are about to discover, I have begun a formal research project testing what is called the spirit guides hypothesis—the serious possibility that spirit guides, including angels, are as real as the light from distant stars.

Why would a responsible scientist reexamine his beliefs about the reality of something as farfetched as spirit guides and actually propose that formal research about them, or with them, is justified? The answer is simple: this is where the evidence is pointing.

For me scientific integrity means following the data where it takes you. It is my conviction that scientists have a responsibility to pursue the path revealed by their emerging data. This may require that we question, if not ultimately reject, fundamental assumptions and beliefs of our society as well as of our professional colleagues—as well as of ourselves!

My willingness to consider investigating the spirit guides hypothesis was stimulated by the fact that more than 80 percent of legitimate research mediums were telling me, often off the record, that they regularly communicated with spirit guides, including angels. Moreover, these "lab rat" mediums (as some liked to call themselves) were claiming that it was their spirit guides that helped them make contact with and obtain accurate information about deceased individuals

Often psychics use the terms *spirit guides* and *angels* interchangeably, though their precise meanings are not identical. Historically, angels have been described as spiritual messengers, typically in the service of the Divine, and have never lived in a biological form. Angels purportedly provide messages of guidance—usually for direction, prediction, and protection. Hence, angels are believed to serve as spirit guides. Some also claim that deceased people can function like angels, providing direction, prediction, and protection, which Susy Smith has demonstrated.

When I first heard the mediums' claims about their purported spirit guides and angels, my instinct was to ignore them. I simply assumed, unfairly, that because mediums as a group are a little weird, by nature they hold bizarre, unfounded, and often foolish beliefs.

It was enough of an academic challenge to be open to the possibility of life after death and test the survival-of-consciousness hypothesis in the laboratory; I reasoned that were I to add spirit guides to the list of spiritual possibilities and begin a research program around this hypothesis, I would probably lose all scientific credibility, even in the eyes of parapsychologists!

The professional and personal challenge I faced, however, was that a lot of the research mediums whom I've worked with, along with their spirit-guide claims, deserved to be taken seriously.

Is it responsible to simply reject what they experience and believe because it is counter to my experiences and beliefs and puts me in scientific jeopardy?

How Spirit Guides and Angels Are Like Oysters

People sometimes experience distaste, if not disgust, when they eat certain foods or hear certain words.

Unless they had a positive experience with these foods or words when they were children, and were raised to develop a taste for them, they might experience them as foreign. Moreover, if in their formidable years, experiences with the foods or words were negative, they might have developed a strong aversion.

I do not like raw oysters—or cooked ones, for that matter. They look slimy and squishy, and I can literally feel sensations of gagging and nausea if I imagine eating them. My parents did not like oysters; they never served them at home or ordered them in restaurants.

Meanwhile, a close friend I'll call Claire had been brought up by parents who loved oysters and shared their fondness for them with their daughter. Claire not only savors oysters, but I have on occasion endured hearing her slurp down raw oysters with a smile on her face (not mine).

Claire told me that her reaction to the concept of angels was like what I experience with oysters—an aversion associated with a "weird and icky" response (her description). What is curious is that Claire is a spiritual woman who deeply believes in God but finds the idea of angels fanciful. Her parents believed strongly in God and prayed regularly, but they questioned other spiritual premises including the existence and nature of angels.

What is equally curious about me is that I was raised on the South Shore of Long Island with parents who loved raw as well as cooked clams. Fresh Long Island clams were readily available in supermarkets and local restaurants, and I developed affection for clam dishes as well. Though my taste for raw clams was tainted by an undergraduate course in zoology, which required that we dissect them, I soon forgot the details of their anatomy. To this day I still enjoy baked clams and linguine with white clam sauce.

The fact that I had learned to develop a positive taste for clams and an aversion to oysters, and Claire had learned to develop a positive taste for God and an aversion to angels, illustrates how it is possible for us to develop prejudices and aversions that are based on experience rather than on the intrinsic nature and reality of the substances or concepts in question.

Do you like oysters? Do you like clams? Do you have an affinity for the idea of God? Do you have affection for the idea of angels?

Whatever your current tastes and preferences are for concepts like God and angels, the challenge raised here is for us to carefully reexamine our current attitudes and preferences about the existence and nature of a larger spiritual reality in light of the emerging new evidence. Science should not be about personal tastes—concerning physical foods or conceptual hypotheses; it should be about empirical evidence. I might add that this criterion, what constitutes evidence, might also be questioned to allow for an expanded vision.

First Reasonable Doubts about the Santa Claus Explanation of Spirit Guides

What I am about to share was not part of my formal research when the event happened; most of this information is public knowledge now, though, and I am simply recounting my biographical as well as autobiographic perspective. As it happened, the first person who forced me to reconsider my Santa Claus interpretation of the existence of spirit guides and angels was John Edward. The year was 1998. As I described in *The Afterlife Experiments*, John participated in three mediumship experiments in my laboratory before he became a celebrity.

John does not suffer fools lightly. He is a New Yorker, an Italian, and he works out regularly in the gym. He is a loving and caring man, but also logical and tough. John regularly gets specific and even rare information in his readings; his range of accuracy in laboratory testing was often 80 to 90 percent. Though John does not advertise this fact, he not only believes in but also regularly receives information, including personal guidance and direction, from a set of spirit guides that he calls "the Boys." John was raised Catholic and was taught to believe in the existence of spiritual beings, including saints and angels.

For example, John's decision to write his book *Practical Praying* was apparently inspired by the Boys. In a book review for the *National Catholic Reporter* in 2006, Retta Blaney wrote:

The idea, or "message," for the book came from "the Boys," his spirit guides, while he was on a book tour. "At that precise moment, the phone in my room rang," he writes, explaining that it was his publisher asking if he could talk to him. "I told him to come on by, and I immediately started trying to figure out how to tell him what my spirit guides had just told me to do. I was sure that he'd say it was a crazy idea."

I underscore the fact that this is more than John's belief; this has been his ongoing personal experience for many years. In fact, it was supposedly the Boys who inspired John to call me in 1998 and tell me that I should write the book that became *The Afterlife Experiments*. John takes the Boys very seriously.

Though my knee-jerk response would typically have been to dismiss John's experience with the Boys as nothing more than his imaginary playmates—I mean no offense to John here—the truth was that John is too seasoned, strong, responsible, and mature to fit the stereotype of New Age flake. Though I had no idea if the Boys were real or fanciful, my personal curiosity, as well as scientific responsibility, led me to ponder why someone like John would not only hold such a belief, but also regularly claim to have such experiences.

John was too accurate and successful to be dismissed; he deserved being given the benefit of the doubt. For me, the deeply challenging question was not if the Boys potentially were real, but rather, it was how I would determine, scientifically, whether they were real and if they were the source of meaningful information and protection for John.

I had no idea at the time how I might address these or similar controversial questions scientifically.

However, that was then, and this is now. And the opportunity for researching the spirit guides hypothesis is upon us.

GARY E. SCHWARTZ, PHD

Angels on Call

Some mediums and psychics are forthright about the role of spirit guides and angels in their lives; a few of them, such as Mary Occhino, even advertise this belief. I have already presented an example of Mary's remarkable abilities as a research medium. I now turn to Mary's defining belief and experience concerning her relationship with her spirit guides and angels.

At the time this book was written, Mary was the host of a highly successful three-hour-a-day, five-day-a-week call-in radio program on Sirius XM Radio titled *Angels on Call*. Mary chose the title because she was absolutely convinced that her success as a medium, a medical intuitive, and a psychic counselor was due to the fact that her angels were literally on call with her.

For more than two years, I have been doing a weekly half-hour science segment on *Angels on Call*. From time to time, Mary would invite people to call in and share personal experiences of guidance and protection that they believed involved spirit guides and angels. Mary claimed on numerous occasions that not only were her spirit guides and angels interested in our research, but also that they were ready to serve as spirit subjects in the laboratory!

It's one thing for a highly gifted medium to claim that her abilities are facilitated by the active collaboration of her angels. It's quite another for the medium to assert that her angels are willing to be tested to prove that they are indeed responsible for the medium's success. And this wasn't an idle claim; she was expressing it to someone who tests such claims continuously.

Talk about chutzpah or guts (or foolishness)—not only on Mary's part but on the part of the purported spirit guides and angels as well. And like John, Mary is too accurate and successful not to deserve to be given the benefit of the doubt.

If Mary and her angels are willing to participate in research, it seems that they should be given the opportunity.

The Medium-Angel Connection

On a few occasions, I have conducted exploratory personal investigations to see, for example, if a medium could read the mind of the experimenter. In one investigation, I tested a medium (who chooses to remain anonymous) to see if he could distinguish when the experimenter was thinking about someone who was alive versus someone who had died.

As part of the private testing, the medium was requested to identify the sex, approximate age (young or old), and status (living or deceased) of the person the experimenter was thinking about; I was the experimenter. The investigation was what is called a 2 × 2 × 2 design (half male, half female; half young, half old; and half living, half deceased). To my amazement, the medium was more than 90 percent accurate in making these judgments.

When I asked the medium how he did this psychic feat, he claimed he couldn't do it; what he did instead was asked his spirit guides for the information!

Was this cheating?

I repeated the investigation, only this time I asked the medium to not seek the assistance of his spirit guides and angels, and try to do it himself. Under these conditions, his accuracy in reading the mind of the experimenter decreased to chance.

This proof-of-concept exploratory investigation does not establish the role of his spirit guides and angels in mind reading; the results could have been due to differences in the medium's belief. In other words, if he thought he would fail, he would fail, because this was his belief.

This investigation raised questions: If spirit guides and angels exist, could they provide the medium with information he or she could not normally receive? If spirit guides and angels exist, are they asking to be validated, and do they want to be heard?

Being a die-hard agnostic, I strive to keep an open mind about such possibilities, despite my upbringing, acquired tastes and distastes, and prevailing scientific beliefs.

Are spirit guides and angels real? My intellectual response is, "I don't know. Could be yes, could be no. Show me the data; I'm open." My emotional response is more akin to my aversion to oysters than my fondness for clams. I would have been happy if some other scientists were systematically addressing the spirit guides hypothesis, but as far as I know, at the time this book was written, there were no established research laboratories investigating this hypothesis.

So why did I ultimately decide to bring this question into my laboratory? The Johns and Marys of the world, insisting that spirit guides and angels are real, were part of the reason. The other reason was that completely unexpected events happened in both my professional and personal lives to indicate that it was essential that this controversial hypothesis be given a fair and honest opportunity or chance to scientifically prove itself.

If spirit guides exist—whether they are literally angels or merely acting like them in the role of messengers—and if they can play a caring and protective role in our individual and collective lives, then maybe we should open our minds and hearts to what they are saying. If they actually know things we don't, and they can see possibilities that are beyond us, it would be potentially self-destructive to ignore the information.

Given the present predicaments of humanity and the planet, it seems prudent for us to seek all the wisdom we can receive, even if this means developing new tastes—and even if they are squishy going down.

12

PUTTING ANGEL SOPHIA AND
HER INTENT TO THE TEST

It is not known precisely where angels dwell
—whether in the air, the void, or the planets.
It has not been God's pleasure that we should
be informed of their abode.

—Voltaire

I had successfully avoided thinking about spirit guides and angels, at least within the university setting, until one fateful afternoon in the summer of 2003 when a new postdoctoral fellow asked me if I wanted to talk with my guardian angel.

If you typically have trouble with the concept of spirit guides, and in particular have difficulty with the idea of angels, it is worth remembering our discussion that "angels are like oysters" in the previous chapter. You can at least read about angels as I can write about oysters. In the same way that there is big difference between typing the word *oyster* and actually swallowing one, I am not suggesting that you swallow the idea that guardian angels are real. I just encourage you to read about the remarkable set of events that transpired in my life and in my laboratory and come to your own conclusions about them.

Remember, I am reporting what actually happened, changing the names and certain identifying details as appropriate to protect anonymity. The person I was meeting with was in his forties, with a PhD in cardiovascular physiology, who was about to begin a two-year research fellowship at the University of Arizona, which was funded by a grant from the National Institutes of Health. He had selected me to be his primary mentor. Prior to this face-to-face meeting, I had spoken with him once on the telephone. As you can probably surmise, in that conversation he had not mentioned anything about spirits or angels. To protect his anonymity—he prefers that his angelic awareness not be widely known—I will call him Dr. Jackson.

Though this was in no way related to the topic of his postdoctoral fellowship, Dr. Jackson was aware that I had conducted research on life after death. He confessed that he could readily see spirits, including angels, and he claimed that one of my angels—a female—was standing in my office behind my right shoulder!

In my thirty-plus years of holding research meetings in various laboratories at Harvard, Yale, and the University of Arizona, I had never heard a claim about an angel in my office, let alone "my" angel. Yes, some mediums had claimed to see deceased spirits in my office and laboratory, but then they regularly reported seeing dead people in houses, restaurants, and university offices.

Though I was intrigued by Dr. Jackson's angel statement, to say the least, I did not feel it appropriate to explore his purported extrasensory perceptions at that time. Our task in that meeting was to map out plans for his research fellowship, not to hold a conversation with my supposed spirit guide. However, in light of other events related to angels that were cropping up at that time—for example, I had been curiously gifted a book about the history of angels, which I had not yet read—I told Dr. Jackson that I would look forward to talking with him about angels, off the record, at a future time.

As it turned out, Dr. Jackson was going to join me and two other colleagues in a daylong trip to a research clinic evaluating a purported energy-medicine device. On the way to the clinic, I asked Dr. Jackson if he was willing to share some of his early as well as current experiences related to angels.

It appears that angels had played a central role in his life just as they did in Mary Occhino's. Dr. Jackson even said that his earliest childhood memory involved being with one of his lifelong angels. He claimed that because of his intimate relationship with them, he was sometimes able to perform healings on family members and friends.

Dr. Jackson believes that everyone has angels that assist throughout life, that most of us have no awareness of guardian angels, and that the majority of us are blind to the intrinsic value of coming to know our angels and consciously working with them.

Dr. Jackson felt that it was safe to share this information with me. He claimed that his spirit guides were instructing him to awaken me to existence of angels, including the specific ones associated with me. Of course, I entertained various alternative possibilities, including that Dr. Jackson was delusional, that he was pulling my leg, and that he might even be a secret Randi plant—with the Amazing Randi, you never know what's up his magician's sleeve. However, as far as I could tell, Dr. Jackson was a successful, responsible, and highly recommended PhD scientist who just so happened to hold strange beliefs similar to those of such mediums as John Edward and Mary Occhino.

GARY E. SCHWARTZ, PHD

Angel Sophia's Sudden Appearance

After returning to Tucson and before I went to sleep that night, I wondered whether it was possible that I, and everyone else, actually had one or more guardian angels. I realized that unlike the survival-of-consciousness hypothesis, which was meaningful only to individuals who have experienced the loss of a loved one and have a personal reason to care about life after death, the spirit guides hypothesis had the potential to be meaningful to all of us. If we all could potentially improve ourselves by becoming increasingly aware of higher spiritual guidance, the implications for improving our daily lives made the spirit guides hypothesis more appealing.

Starting with when I was a professor at Yale, I would from time to time "ask the Universe" a question, and novel thoughts would typically pop into my mind that often could be verified.

The first such question I asked of the Universe was, "Could you give me another name for God?" What immediately popped into mind was the name Sam. When I first heard "Sam," I couldn't help laughing. I wondered, was the name Sam a product of my creative unconscious, or was I possibly living in a Woody Allen movie? However, when I looked up the origin of the name in my Webster's, I got Samuel. I discovered to my astonishment that Samuel comes from the Hebrew Shemuel, which translates literally as "the name of God."

After carefully considering nine possible conventional explanations for why this name might have popped into my mind, I seriously entertained the possibility that while asking the Universe a question in a state of deep authenticity and genuineness, it had somehow provided me with a concrete answer that I could later verify. In *The G.O.D. Experiments* I confessed that I initially was reluctant to explore this potential hypothesis; in fact, I avoided asking another question of the Universe for more than a decade. However, when I returned to asking such questions of the Universe, the good answers came with sufficient regularity for me to no longer, with integrity, dismiss this avenue of inquiry.

So that night, after returning to Tucson, I decided to ask the Universe, "Did I, as Dr. Jackson claimed, have a female angel, and could the Universe show her to me?"

What happened next was unique for me—I got absolutely nothing: no names, images, feelings, or memories. The complete absence of any subjective experience took me by surprise. I went to sleep impressed with the complete failure of my request.

The next morning I happened to read an article in the magazine *Scientific American* about so-called censor genes. These are genes that suppress certain genetic potentials from being expressed. It occurred to me that I had been so well conditioned to believe that angels were like Santa Claus—a playful fiction and nothing more—that my mind was probably censoring any awareness of a potential angelic presence in my life, if indeed there was one.

That night, I decided to try my personal experiment again. I consciously attempted to release any mental censorship I had about angels. I asked that my mind be opened to all possibilities, and then I did something new. Instead of asking the next question of the Universe, I directed my question to my possible angel instead.

I said in my mind, "Angel, if you are here, I would love to see you and learn your name."

What happened next was totally unlike anything I had ever experienced before or since that night. Whereas I rarely see images—I mostly think in abstract terms; even my dreams, when I remember them, are relatively flat and colorless—I experienced the appearance of a large glowing figure hovering above the foot of my bed. My bedroom has high ceilings. The spirit, or hallucination or whatever it was, looked to be at least eight feet tall.

The spirit appeared to be female, with flowing blond hair. Around her shoulders were bright lights that looked to be in the shape of wings, but they could have been reflections off her arms and body. She seemed to be wearing a whitish dress. She was smiling. She appeared to be loving and gentle, yet strangely powerful.

I asked her in my mind what her name was, and what popped in was the name Sophia.

At that moment, my rational skeptical mind kicked in— I thought something to the effect of "this can't be real"—and the vision of the female spirit vanished. Just like that, poof, and it was gone.

I realized that the name Sophia in Greek means "wisdom." I had never heard of an angel named Sophia. Hearing this name for an angel initially seemed as bizarre to me as the name Sam for God was almost twenty years earlier.

The next morning I checked to see if there was an angel named Sophia on angel websites. I Googled the words *Angel* and *Sophia*, and what I uncovered left me breathless. There were more than five million entries for Angel Sophia, and some of websites revealed a profound set of religious beliefs associated with Angel Sophia.

Depending upon the internet source, Angel Sophia was described as being one of the following:

1. The first emanation of the Divine and the Mother of all creation, including of all the archangels

2. The feminine expression of the Divine—the male expression of the Divine being an angel purportedly called Metatron

3. The wife of Metatron

There were also many references to writings about something called the "Pistis Sophia" and a Christian denomination focused on her.

First there was Sam, the name of God, and then Sophia—two relatively unknown big names from religious history. As I had done with the name Sam—I had asked more than fifty staff members, students, and faculty if they knew the origin and meaning of the name Sam and only one person knew, meaning it was not common knowledge—I asked a cohort of people who actually knew something about angel lore if they had ever heard of an angel named Sophia.

The first nine people I asked included Dr. Jackson and the person who had gifted me the *History of Angel*. None of them said that they had heard of an angel named Sophia.

However, the tenth person I asked, who happened to be visiting me from California and was an ordained minister as well as a successful corporate executive, not only knew about Angel Sophia, but a close associate of his was a scholar who had just completed a book about Pistis Sophia. I had to wonder if this was a synchronicity. When he heard how I came to discover Angel Sophia, he was inspired to contact his friend, who sent me an inscribed copy of his book.

Try putting yourself in my shoes at that time: You had a seemingly innocent meeting with a new postdoctoral fellow, and in the course of the meeting the scientist claimed that a female angel was standing behind your right shoulder and wished to speak with you.

You eventually mustered the courage to attempt having some sort of personal angel experience yourself. You had a surprising vision of a glowing white, angelic-like female and heard the name Sophia, which you initially presumed has nothing to do with angels.

You then discovered in angel lore that there really was one named Sophia, and that she was viewed in some circles as the mother of all angels.

You then conducted a small informal survey with people who knew something about angels, and none of them said they had heard of an Angel Sophia. However, the tenth person not only knew of her existence, but he even knew a scholar who had just written a book about Pistis Sophia! Talk about a chain of synchronicities.

What would you have done with this information, especially if you knew mediums like John Edward and Mary Occhino who, like Dr. Jackson, were convinced that angels were real? Dismiss the information? Run from it? Put your head in the sand?

I might well have done this, except my mind wouldn't let me. What unfolded was a new proof-of-concept personal experimental test.

Putting Angel Sophia to a Test

I should confess that I have what some people call a disconcerting, if not bad, habit, especially at a university. I sometimes jokingly tell colleagues and even strangers that as a scientist I have developed a "disease called science."

It is more than L.O.V.E.; it is a habit that some would say can be likened to a dependency. What I mean is that in an automatic and often uncontrollable way, when I hear someone share a belief or an experience—the L in L.O.V.E—my mind effortlessly does the following:

1. It converts the person's belief or experience into a question.

2. The question is then turned into a hypothesis.

3. The hypothesis is operationalized, meaning the hypothesis is refined so that it can be potentially measured—the process of 1–3 reflects the O in L.O.V.E.

4. The operationalized hypothesis is then transformed into an experimental design.

5. I will typically conduct the experiment in my head as an Einstein-like thought experiment.

6. And then, if it is feasible—meaning I have the time, funds, equipment, and so forth—I will feel the strong desire to conduct the experiment. This is the progression from O to V— the verification in L.O.V.E.

Because this process is so automatic and effortless, I often do not think much about it, unless the hypothesis happens to be exceptionally controversial and it turns out that it is actually possible for me to test it—be it an exploratory investigation or a full-fledged university experiment. This is what happened following my vision and upon becoming aware of beliefs about Angel Sophia. I realized that I had the possibility to conduct an exploratory investigation testing whether some sort of spiritual being with the name Sophia was somehow connected to me. The question I had asked was whether I was brave enough—some might say foolhardy enough—to actually conduct such a proof-of-concept investigation.

As fate would have it, the opportunity to conduct such an investigation in my personal life was dropped in my lap, so to speak. What transpired forever changed how I viewed the potential existence of a larger spiritual reality and our ability to investigate and learn from it.

It occurred to me that if Susy Smith was actually watching over me, then she would probably know about my seemingly anomalous experience with a purported angel named Sophia, and she would want to help me (1) verify if Sophia actually existed or (2) establish that Sophia was a figment of my imagination, whichever was the case. At this point I had no idea.

If Sophia was my angel—she might also be an angel to many others as well, if she really existed—and she was willing to appear in my bedroom, then she would most likely know that I was a person who suffered from a disease called science, and that I would want to determine, experimentally, whether she actually existed.

Moreover, if Sophia had been watching me for a long time, she would know of my relationship with the late Susy Smith, not only when Susy was in the physical but also after Susy had transitioned to the other side. I also speculated that if Sophia had my best interests at heart, she would likely be willing to collaborate with Susy to help validate her existence. Otherwise, why would she show up in the first place?

The idea popped into my head that in principle, Susy might be able to bring Sophia to an appropriately receptive medium. I wondered about the possibility of a deceased spirit actually bringing an angelic spirit to a medium. And if Susy succeeded in bringing Sophia to a medium, would the reading potentially verify the angel's existence?

At this point I was simply doing a thought experiment. I realized that if an exploratory, proof-of-concept investigation were actually conducted with positive results, the possibility existed for doing formal, controlled, and systematic research on the spirit guides hypothesis. But before addressing the question of whether alleged spirit guides could provide meaningful information for people individually and collectively, it was essential to determine scientifically if spirit guides existed, period. This was where I began.

A Disconcerted Medium's Susy-Sophia Reading

As it so happened, I was scheduled to give an address presenting our latest energy-medicine research—funded at the time by the National Institutes of Health—at a conference that was being held not far from the home of a gifted research medium. To preserve the requested anonymity of the medium, I have changed his name as well as the name of the city where the reading took place. We will call the medium Harry and the city Baltimore.

Harry lived about an hour from where the conference was being held. He had previously provided exceptionally accurate information regarding Susy Smith; moreover, he claimed that Susy would spontaneously visit him from time to time. But it turned out that Harry was one of the few mediums I have worked with who neither believed in, nor claimed to communicate with, angelic beings. Harry was a fairly down-to-earth person who was known to enjoy social gatherings as well as alcoholic spirits—I know many mediums who understand how to enjoy liquid spirits as well.

I deeply respect Harry's skills as a medium; the fact that he had little use for the idea of angels made his selection for my private exploratory investigation all the more interesting. So I called him a week before the conference and asked if I could have a personal reading with him. Harry happened to have some free time after my conference was over, and we scheduled the appointment.

Technically, what transpired was not the double-deceased version of the spirit-mediated model since Sophia presumably was not deceased; as an angel, she had never lived in the physical. However, it was a deceased-mediated paradigm since Susy supposedly brought Sophia to Harry.

When I got to Harry's house, after expressing greetings and sharing personal kinds of catch-up information, Harry led me into the room where he conducted his readings. Though I had tested Harry in laboratory research on numerous occasions, I had never had a personal reading with him. I told Harry that I wished to hear from two beings in spirit. The first I would identify by name, the second would presumably be brought by the first, and I would not provide any information about this individual. At no time did I say, or imply, that one of the spirits was potentially an angel.

I then told him I wished to hear from Susy Smith. He was pleased to comply with this request. He shared information, purportedly from Susy, for approximately fifteen minutes. Though this information fit Susy well, most of it was scientifically useless because: (1) I had given him Susy's name, (2) he had previously participated in research involving Susy, and (3) he had subsequently read some of Susy's books. Since Susy was not the focus of this exploratory investigation, these factors were unimportant.

Next I told him that I had asked Susy to please bring along another individual in spirit, and that I would like it if Harry did a reading with this individual.

What happened next was thoroughly confusing to Harry, and a bit upsetting for him as well.

Harry reported seeing a very tall woman with blond flowing hair.

He described her as radiant, bigger than life, and sort of floating above the floor.

He said that he could not look at her eyes; she was very powerful.

He said he felt inadequate in her presence. I have never heard a medium speak this way of a deceased person whom he or she has contacted.

Further, Harry could not ascertain where she had previously lived on earth or how she had died.

He sensed she had been in Spirit a very long time, had an S-sounding name, and was somehow connected to me, but he could not figure out how.

He said she gave no indication that she was a blood relative, or even a personal friend of mine, yet there was a very strong emotional and intellectual bond between us that made no sense to him.

Harry not only felt inadequate in terms of his mediumship skills with this individual, but he said he even felt embarrassed to be reading her.

He said he had never experienced this hesitancy, and I had never heard a medium say that she or he felt embarrassed reading a particular spirit.

Save for my confirming that I did wish to hear from a woman, and that she probably would appear to have flowing blond hair, afterward I did not give Harry any additional feedback about his potential accuracy or lack thereof. Moreover, I told Harry that for personal as well as scientific reasons, I could not at that time give him any indication about who the woman might be or why I had requested that Susy specifically bring her to the reading.

After thanking him greatly for his time and efforts, I left him in a rather befuddled state. While driving back to town in my rental car, it began to dawn on me that what had just transpired was possibly unique in the history of afterlife, death, and spiritual research.

Susy Smith, a distinguished deceased author of thirty books in parapsychology and the apparent inventor (from the other side) of the double-deceased paradigm, had just been read by a gifted research medium. The medium claimed that Susy had brought to the reading another woman in spirit whose glowing presence sounded very much like what I knew of Sophia, the angel.

Had Susy, a deceased woman, brought an angel, in this case Sophia, to Harry?

Had Harry read an angel and not known it?

Was Harry's personal discomfort, coupled with his inability to get standard information about the woman, like how she died, potential evidence that this spirit was indeed someone who had never lived on the earth, and therefore had never died?

Obviously, a single reading does not make for a definitive scientific test—in academic research, this would be called nongeneralizable data (you may recall our discussion of the federal government's definition of research in chapter 1).

However, single readings can serve as beacons of opportunity, revealing what can potentially be explored and documented systematically in the laboratory.

What was I to do now?

The Birthing of the Sophia Project and Its Umbrella Voyager Program

The more I thought about what had transpired with Harry, the more I realized that experimental research on the spirit guides hypothesis was possible in principle. However, I had no research funds for conducting such experiments; moreover, I had no idea how to raise such funds. I also realized that even if I had funds for such experiments, it was professionally dangerous for me to conduct them. As mentioned previously, the afterlife research was already challenging my scientific and academic credibility.

Talk about going from the frying pan into the fire. I seriously wondered if I began conducting research on the spirit guides hypothesis whether I would risk the possibility of being fired from the university and metaphorically thrown in the academic dungeon. Though I had been a tenured professor at both Yale and the University of Arizona, I was well aware that the scope of academic freedom protected by tenure had its limits and boundaries, and I was potentially close to crossing them.

Similar to my first personal experience with asking the Universe questions, which I intentionally suppressed and successfully avoided exploring for over a decade, I avoided exploring angel-related questions for a few years—except for one more secret feasibility laboratory investigation, described in the next chapter—until a strange, purportedly angel-initiated opportunity presented itself. It appeared that Sophia is just a willful as Susy.

I wondered, was I being assisted, if not tested, by higher spiritual beings? In brief, here is what happened.

In the spring of 2006, a businesswoman I will call Suzanne contacted me through a third party, a businessman whom I will call Ed. Ed claimed that he represented a successful woman who wished to make a private donation to me and the work.

When I first spoke with Ed, he stated that Suzanne did not care what I did with the money—I could give it to the university and use it in the laboratory, or I could deposit it in my checking account and use it personally for the work. Also, Ed claimed that Suzanne did not care what topic I researched; the choice was entirely mine.

I was immediately suspicious. My questioning agnostic mind quickly ran through the spectrum of possibilities, including that this businesswoman was crazy, that Suzanne and Ed were teasing me, and/or that they were potentially secret agents of a superskeptic seeking to discredit the work. However, a little detective work conducted by my administrative assistant revealed that both Ed and Suzanne appeared to be respected and seemingly sane members of the greater Tucson business community. I agreed to meet them to explore Suzanne's wishes.

Suzanne and Ed turned out to be interested in matters of spirituality and healing and as the conversation unfolded, Suzanne made an extraordinary, off-the-record confession. She said that she had spirit guides and regularly communicated with them.

Suzanne then claimed that a few months earlier, while she had been visiting Flagstaff, Arizona, her guardian angels told her that she should help support Dr. Schwartz and his research. I looked to Ed for confirmation; was I hearing Suzanne correctly? Ed nodded his head.

Suzanne said that her angels told her that she should take $10,000 of the profits from the recent sale of a real estate property and donate it to me, no strings attached. Suzanne claimed that I would know what part of the work most needed the funds.

Over the years I have met numerous potential donors, and some have made seminal contributions to the work. However, no prospective donors had made the extraordinary claim that they were being instructed by their angels to make a contribution and that I should use the funds any way that I wished!

I asked Suzanne if she was aware that I had a personal interest in the possibility of conducting angel research, and she said no. For the record, at that time only a few of my closest colleagues were aware of my new interest in researching the spirit guides hypothesis. Moreover, I had not told anyone, including the medium who participated in the Susy-Sophia reading, that I was pondering the feasibility of testing this hypothesis using research mediums in a laboratory setting.

Suzanne went on to claim that she had been following my mind-body-spirit research and writings for more than ten years. She said that she had attended a number of public lectures I had given over the years, and she had watched some of the television documentaries about my work. She further said that if I would sign her copy of *The G.O.D. Experiments* book, she would double the donation for a total of $20,000!

I was frankly stunned. I had never heard the claim that angels could, let alone would, direct someone to support research about them.

I wondered whether the prospective donor was deceiving not only me but Ed and even herself as well. Or was it possible that as John Edward and Mary Occhino were claiming, spirit guides and angels were insistent that it was essential for humanity to wake up to their existence, and that it was time for experimental angel science to begin?

As I listened to Suzanne's seemingly earnest confession in the presence of Ed, her confidant, it occurred to me that the only way I could conceivably accept such a gift would be to use the funds to create a small university research project to test the underlying premise of the gift itself. In other words, the research project would test if (1) spirit guides, including angels, were real, (2) if they could provide guidance and direction, and (3) if this information could include guidance and direction concerning the conduct of spirit guide research.

I told Suzanne and Ed that I needed to think about her gracious offer, and that I would like us to meet again in a few days to consider a formal proposal. Even though Suzanne said that I could use the funds in any way I thought benefited the work, I explained that I wanted her and Ed to examine and potentially approve what I proposed. I further explained that I was going to request something that would honor the potential reality of what Suzanne had confessed; I would ask her to please confer with her spirit guides and angels, and see if they agreed with my plans for the funds.

We met a few days later. I shared with Suzanne and Ed that I could not, with integrity, simply accept her gracious donation based on her belief that she was following the guidance and direction of her angels. I explained that as a scientist, I did not know whether spirit guides and angels existed. However, I then told her of my set of experiences with (1) meeting Dr. Jackson and his claim of seeing one my guardian angels with me in my office; (2) my personal attempt to discover whether there was an angel connected to me; (3) my uncovering evidence on the web for the historic claim that there was an angel named Sophia—neither Suzanne or Ed had ever heard of an angel named Sophia; and (4) the encouraging results of the private proof-of-concept feasibility investigations that I had conducted to see whether it was possible to bring the spirit guides hypothesis into the laboratory.

As you can imagine, Suzanne and Ed were somewhat surprised as well as grateful. Suzanne explained that although she and her angels were not interested in being lab rats personally (and for experimental and ethical reasons it would not have been appropriate to include them), they agreed that it was time for scientists to begin formal research on the spirit guides hypothesis.

I proposed that her unsolicited donation be used to initiate formal research within the university on the potential reality of spirit guides and angels, and that we name it the Sophia Project. I further proposed that the first research explore the inspiration for the project: the premise that spirit guides and angels could communicate with people and provide guidance, direction, and sometimes protection. The plan would be to conduct a university Human Subjects Committee IRB-approved structured-interview experiment with a representative sample of people working across the United States and who functioned as professional mediums and angel communicators.

I suggested that the interview be designed to systematically explore the personal experiences of knowledgeable professionals who regularly communicated with deceased spirits and angels. And in the process of conducting the interviews, we would request that they—both the professional communicators and their purported spirit guides—offer specific suggestions for how best to design subsequent research experiments. The purpose of the interviews was not to evaluate the validity of their experiences but rather to explore in some serious depth what their experiences were. This seemed like a responsible and relatively safe place to begin the research.

In the summer of 2006, the Sophia Project was born. I hired a research assistant to work on it one day a week. It took us a year to (1) design, (2) test questions on each other to make sure they were clear, (3) submit the necessary documents to the university, and (4) ultimately receive approval to conduct the structured interview protocol. At the time this chapter was written, the interviews were almost completed, and here are a few preliminary responses from the communicators.

What kinds of information do deceased people give you?
"They are okay; they are around the living, [giving] advice for the living . . . resolving issues."

"Whatever communication needs to happen so that healing can take place—healing for both the dead person and the sitter."

"Messages of love to the people I'm working with . . . 'I'm still around,' 'I'm helping you out.' It is coming from a place of love that wants to be expressed. They want to correct any unresolved issues."

What kinds of information do angels give you?
"Uplifting messages, healings, help with problems."
"It is supportive information to help someone find answers for spiritual and personal growth or healing. All information is going to make a positive impact."
"It's always about creating the best possible life experience ... about facilitating the process that allows people to live their bliss."

Meanwhile, thanks to Canyon Ranch and a few visionary donors associated with the ranch, funding was eventually provided for a larger program of research called the Voyager Program. The name was suggested by Dr. Jonathan Ellerbe, spiritual program director at Canyon Ranch and author of *Return to the Sacred*. This umbrella research program was created to conduct "research integrating energetic and spiritual mechanisms in healing and life-enhancement." Research on the spirit guides hypothesis in the Sophia Project falls within this broader program of questions and applications. Some of the research you are reading about here was made possible by the Voyager Project's essential support.

Suzanne's experience reminds us, once again, that what we are exploring in this book is not only research in the scientific laboratory, but research in the laboratory of our personal lives as well. What we are witnessing is the increasing application of the scientific method for the enhancement of our ability to both discover and manifest human possibilities that many of us—especially academic scientists—are not only unaware of but that we presume are unreal and impossible.

If there is one lesson emerging from my Sophia experiences, it is the need for humility with integrity. The science of the seemingly impossible is about to get even stranger. Acquiring new tastes may become desirable and even necessary.

13

DETECTING THE PRESENCE OF ANGELS IN THE BIOPHOTON LABORATORY

Angels fly at light speed, because they are servants of the Light.

—Eileen Elias Freeman

Throughout recorded history, it has been written that spirit guides, including angels and the Divine itself, are associated with light. In fact, the Divine is often called the Light. The very nature of human understanding itself has also been associated with light. We speak of being "enlightened" about something, and when we discover the truth, we are "seeing the light."

When I was in New Haven, I had five Yale chairs in my office that prominently displayed the university's Latin motto, *Lux et Veritas*, meaning "Light and Truth." The presence of these striking Yale chairs regularly reminded me about the special nature of light and its relationship to knowledge and wisdom.

When I use the metaphor of our personal energy, and even the purported energy of spirit guides and angels, being like the light from distant stars, I do so not simply for metaphorical reasons but for serious scientific reasons as well. For if any aspect of contemporary physics deserves the term *spiritual*, it is our understanding of the nature of particles and waves of light.

The Wondrous Nature of Light and Angels

Much has been written about the remarkable, counterintuitive, and mysterious nature of light. Though contemporary physicists know much about the properties and behavior of light, they have little understanding about the essence of light—in other words, what it is that enables light to manifest its weird and wondrous properties, what physicists playfully term "quantum weirdness."

Here are a few well-established counterintuitive properties of the nature of light and their curious similarities to long-standing beliefs about spirit guides, especially angels:

1. **Light is a particle and a wave:** Light sometimes behaves as if it were a particle—localized in space—and at other times as if it were a wave—distributed in space. There is the classic single-slit/double-slit quantum physics experiment that shows light acting as a particle in the first case and as a wave in the second. Some physicists believe that light is neither a wave nor a particle, but rather a so-called wavicle—an idea that is virtually impossible to imagine.

In a deep sense, light is not a thing and does not have a specific shape. It is curious that angels are purportedly able to appear in different forms, and they also can appear either localized or in multiple places at the same time.

2. **Light is virtually massless:** Light is typically described as massless, or weightless, especially when it is functioning like a wave. However, a minority of physicists speculate that a photon of light might have a tiny amount of mass when it is functioning like a particle.

Angels are presumed to be virtually translucent, which implies being weightless, in their normal, spiritual state.

3. **Light travels at a fixed speed:** Light is presumed to travel at a fixed speed in the vacuum of space—approximately 186,282 miles a second—regardless of the speeds at which objects are traveling toward or away from it. So, no matter how fast we travel—even at the speed of light itself—we can never catch up to a specific particle or wave of light traveling in space, because it will always be moving away from us at this speed.

Hence, light can always travel faster than we can. So too, it is said, can angels. This may be why, if we catch a glimpse of them, they seem to so readily disappear and reappear.

4. **The way light behaves when it is entangled:** If two photons of light are entangled or have identical properties, such as the same spin, and the spin of one of photon changes, then the spin of the other change instantaneously, even if they were separated by trillions of miles. In other words, the two photons will behave as if there is no distance between them. Moreover, many physicists speculate that no faster-than-light-speed communication has traveled between them. Rather, the observed

association is believed to be completely timeless and utterly instantaneous.

It has been claimed that angels and their actions, especially healings, can also function instantaneously, since they are supposedly eternal or timeless, and are believed to melt with or be entangled with those they heal, as if to change or elevate their spin.

5. **Light frequencies are mostly invisible:** What is possibly the most implicitly spiritual factor about the nature of light is that with the naked eye, we see only the tiniest fraction of the spectrum of all light frequencies in the Universe. Imagine the full electromagnetic spectrum to be the height of the Empire State Building, or 1,454 feet tall. The portion of what we could see would be much tinier than a layer of paint, or less than one billionth of the known frequencies of light in the Universe.

Simply stated, we are literally blind to most of the light frequencies present in the Universe. Angels are presumed to be energy beings vibrating at higher/faster frequencies. We can't see radio waves (lower frequencies) or gamma rays (higher frequencies) with our naked eyes, and we accept this fact. It seems prudent that we keep an open mind to the possibility that for the same reason we cannot see higher-frequency cosmic rays with our unaided eyes, we cannot see higher-frequency angels/beings of Light, at least under normal circumstances.

One wonders what might be discovered if we used state-of-the-art digital camera systems that detect higher frequencies of light and searched for the potential existence of higher-frequency beings.

6. **When light is super dim:** Finally, it is now well documented that even in the tiny sliver of light frequencies we can see with

our eyes, the intensity of the light may often be too weak for us to experience it consciously. Optical telescopes can detect the faint light of distant stars because they magnify the light intensity. And while retinal cells in the eye are actually so sensitive that they can register single photons of light, their neural signal is too weak for us to experience consciously.

Angels are often assumed to be present in weak intensity states, though exceptions have been claimed, and as mediums tell us, angels are constantly around us, filling us with their energy, guiding and protecting us, but we never detect their presence—not consciously, at least. One wonders what might be revealed if we used state-of-the-art supersensitive camera systems that can detect and quantify single photons of light and searched for the potential existence of weak-light-intensity spiritual beings.

This is just a sample of the remarkable properties of light that physicists have discovered in the twentieth century. What I find especially curious is that for thousands of years some mystics, who claimed to have spent time with angels, have attributed a set of properties to them that were only recently discovered by physicists to reflect the nature of light as well.

Is this merely a coincidence, or is some deeper truth being revealed?

Measuring the Effects of Humans and Angels on Plants

In *The Energy Healing Experiments*, I describe a series of experiments my colleagues and I conducted using state-of-the-art supercooled low-light digital CCD cameras. This class of computer-controlled cameras is typically used to measure the light from distant stars and galaxies in astrophysical research. We adopted this innovative technology to measure super-weak intensity-coherent light spontaneously emitted by all living systems, including animals, plants, and single-celled organisms.

The camera we initially employed in our energy healing research was cooled to minus 95 degrees centigrade and cost $100,000 new; the cameras we used for our current angelic and related spirit-presence research included one that was cooled to minus 75 degrees centigrade and was somewhat more affordable; it cost $30,000 refurbished. Both cameras can register single photons of light on their supersensitive CCD digital arrays; both can be programmed to take long-exposure pictures in completely darkened spaces (in light-tight chambers and rooms) over minutes or even hours. And both cameras are spectacular in terms of the light images they can capture.

When it became clear that we could use these cameras to measure the invisible light generated by both plants and people, and that the intensity of this light was sensitive to many factors, including the intentions of healers to influence the plants, I began to wonder if the cameras might be sufficiently sensitive to detect the purported photonic nature of angels, since they are supposedly beings of Light.

With my adventurous and creative colleague Dr. Kathy Creath, a research professor of optical sciences at the University of Arizona who holds two PhDs—one in optical sciences, the second in music—I decided, somewhat on a lark, to explore the feasibility of using this technology to measure the possible presence of angels.

We ran four experimental sessions over the course of four weeks, on our own time; the only expense was the cost of the plant specimens themselves, paid for out of pocket. We originally used the $100,000 camera system located in a research laboratory at the university. We were the experimenters; there were no human subjects involved.

A research medium, who also supposedly communicated with spirit guides and angels, and prefers to remain anonymous, claimed that two of her angels could have different effects on plants: one increasing the light emissions, the other decreasing emissions. The angel communicator further claimed that if we invited the two angels to enter the chamber, one on each side, that we might see a difference in the biophoton emissions detected by the camera.

For each of the four evening experimental sessions, Dr. Creath and I placed four germanium leaves, as well as four geranium flowers, in a completely dark, light-tight chamber that was housed in a light-tight darkroom. The computers were in a separate room. The four leaves and their respective flowers were separated by cardboard dividers. After collecting thirty minutes of baseline data, we invited the respective angels to enter the light-tight chamber for thirty minutes. For two of the sessions, Angel X, whose effect purportedly increased emissions, was to be on the left side; Angel Y, whose effect supposedly decreased them, on the right side. The order, determined by coin flip, was switched for the other two sessions.

We had previously documented that leaves typically generate more light than flowers; however, I wondered if the flowers might be more sensitive to the purported differential effects of the angels' energy than the leaves. I included the flowers because of the long history of mythology connecting angels with flowers.

Since this was an exploratory proof-of-concept investigation, we were not concerned with attempting to verify the purported presence of the angels or establishing that they had actually produced the effects. Moreover, the design of the investigation allowed for the possibility that the plants might simply be responding to our thoughts about the possible outcome of the explorations.

We knew that future double-blind studies could be conducted—that is, presuming that the exploratory investigation suggested an approach that was sensitive enough to the subtle energies involved to reveal anything. To our utter surprise, the measurements over the four sessions suggested that something significant was happening.

There were a total of sixteen flowers imaged over the four sessions, with eight of them whose light would supposedly be brighter when compared to the light from their respective rows in control sessions. While from a statistical point of view this is a relatively small sample size of images, when appropriate statistics were performed (such as nonparametric statistics for small samples), the difference was statistically significant: $p < .03$. Though the leaves, on average, showed the same overall effects, the results did not reach statistical significance, probably because of the small sample size.

I later analyzed the flower images from this experiment using a sophisticated image analysis technique called Fast Fourier Transformation (FFT). FFTs are also referred to as spectral analyses and are routinely conducted in acoustic, biophysical, and biomedical research. The mathematical computations take a complex pattern of frequencies and break them down into individual frequency bands for pattern analysis.

FFTs are routinely performed on the ever-changing frequencies of the earth's electromagnetic fields as well as the brain's. They can also be performed on image data. The analyses I performed and graphed used ImageJ software; this image analysis system is provided free of charge from the National Institutes of Health for biomedical researchers and can be downloaded from the Web.

We observed that Angel X was associated with increased ripples or wavelike structures in the patterns when compared with Angel Y. We also found that Angel X was associated with more complex FFT patterns than Angel Y. As you will soon discover, this FFT observation would be replicated in numerous experiments, and it would ultimately reveal some startling properties of apparently angelic and even purportedly higher-frequency emanations of light.

Inviting Angel Sophia and the Light of God into the Laboratory

As intriguing as these findings were, I could not at that time imagine pursuing them systematically. The problem was twofold: First, I had no research funds to conduct a formal series of experiments. Second, I was not in the position to share these seemingly positive findings with the public. At that time I had not yet met Suzanne, or created the Sophia Project or the Voyager Program.

However, the Voyager Program, which investigates the integration of energetic and spiritual mechanisms for healing and life-enhancement—it became the umbrella program for research including the Sophia Project—provided both the funds and the rationale to apply the biophoton imaging system to what I called the Presence of Spirit experiments.

As mentioned in part III, experienced energy and spiritual healers claim that they regularly invite their spirit guides, angels, and even the Sacred (also known as Universal Energy, the Source, the Divine Light, and God) to assist them in their treatments. Moreover, people like Mary Occhino and Dr. Jackson regularly connect with their angels to answer questions on a daily radio show or even to conduct routine business like meeting with their postdoctoral fellowship mentor.

The critical question simply stated is: Is anyone, or anything, really there? Are angels and even higher spirits actually showing up in treatment rooms, on the radio, and even in university offices?

If the answer is yes, could we measure their presence using supersensitive technology, presuming that the spirits were interested in being measured and were willing to collaborate in the research?

There was only one way to find out and that was to perform the investigations. At the time the first draft of this chapter was written, we had completed three controlled investigations testing the feasibility of addressing this question in the biophoton laboratory. The first two involved the presence of Sophia, while the third involved the Light of God.

In these investigations, instead of using a dark background, we employed a black-and-white checkerboard-patterned background. This was suggested, according to a purported angel communicator who chooses to remain anonymous, by Angel Sophia. It was claimed that even in the absence of any plants, leaves, or flowers, the presence of Sophia might be detected.

My research assistant, Mark Boccuzzi (working under the auspices of the Voyager Program), was the experimenter, and he collected three sets of images: (1) a thirty-minute prepresence control image, (2) a thirty-minute presence-of-spirit image, and (3) a thirty-minute post-presence image. Though the chamber was pitch-black and it was in a pitch-black room, we expected that the white squares might be faintly seen in the images. While I had created the experimental design, I was not present during the recording session, having to conduct university business elsewhere.

At the outset of the presence-of-spirit image period, Mark read a sentence inviting Angel Sophia to enter the chamber. Supposedly, Sophia had previously agreed through the communicator to participate in the investigation, and she was ready when we were. While the computer collected its data, Mark listened to music on his iPod.

When we conducted the FFT analyses, the resulting difference among the three images was as clear as it could be. We observed that the period when Sophia was supposedly present showed ripple-like waves and a somewhat more complex pattern.

We also observed a faint angel-like shaped being, with its head tilted toward the upper left-hand corner, and its legs pointing in the lower right-hand corner. Might this be like a Rorschach test in which we project what we want to see? Maybe. But what was indisputable is that the FFT analysis clearly showed a difference between the Sophia image and the pre- and post-baseline images—with Angel Sophia's presumed presence being the only variable.

Struck by the clearly visible FFT ripples in the presence-of-spirit periods, we attempted to replicate the investigation, this time including leaves and flowers. I might add that the current camera was less expensive because it has a smaller CCD array; this reduced the area of viewing so that only four leaves and flowers were visible. In this investigation, daisies were substituted for geraniums. Why? Because this is what Sophia supposedly recommended.

First, we observed that the leaves were clearly glowing in the dark, and faint appearance of the checkerboard was again visible. However, when the FFTs were performed, the difference for the Sophia image as compared with the pre- and post-baselines was even more obvious than it was in the first round of investigations. Dramatic ripple-like waves filled the entire FFT image for Sophia.

The power of the FFT analyses is that they can reveal patterns not apparent to the naked eye, just as the camera reveals light that is normally not visible to the naked eye. Again, just because we can't see it in the raw images doesn't mean it's not there.

In sharing these analyses with Mark, we wondered whether other purportedly higher energies might show similar patterns. Mark had learned a form of spiritual energy healing which originated in the East in modern times. People who practice this technique believe that they receive a universal energy, what they term "Divine Light," and that they can serve as a vessel of the Light of God. I had also learned this technique—it was one of five energy healing techniques I studied as part of my research over the years.

I asked Mark if he wished to explore what might happen if he, as he experimenter, invited the Divine Light to enter the chamber. Mark conducted a third investigation where, in the middle thirty-minute exposure, he meditated and invited the Light of God to enter the chamber.

To our amazement, the results were replicated once again—FFT ripples occurred during the presence-of-spirit period. Curiously, of the three exploratory investigations, the most complex of the three FFT patterns during the presence-of-spirit period happened during the Divine Light segment. I wondered if using a human conduit might not have increased either the intensity or focus of the light.

Surprising Patterns in the Photo Analysis of Cosmic Rays

I might add that these supersensitive cameras not only detect light in a frequency visible to the human eye (as well as some infrared frequency light, depending upon the filters used), but they also detect spontaneous gamma ray/cosmic ray bursts at the same time.

You can think of the camera as detecting two different bands of light frequencies simultaneously: (1) normal light frequencies mostly visible to the human eye and (2) superhigh frequencies of gamma/cosmic rays that are not visible to the naked eye. Moreover, these gamma/cosmic rays are hundreds of times brighter than the biophoton light emitted by living cells. The cosmic ray bursts are generated by stars, including our own sun, and pass through our atmosphere as well as matter: roofs, walls, the metal casing of the camera, and into the chamber itself.

Though they appear relatively infrequently, since gamma rays are so much brighter than biophotons, they leave a superbright spot on the image (relative to the biophotons—remember, both intensities are actually very low). Since it is typically assumed that gamma rays appear randomly in the camera coverage during a given experiment, they should theoretically play no systematic role in the overall results. However, since their intensities are so large (relative to the biophotons), it is preferable to statistically remove them using software.

The analyses reported above of the three investigations and their FFTs reflected the raw data that included the sporadic gamma/cosmic rays—for convenience, I will simply call them cosmic rays. This is also the case for the preliminary investigation I performed with Dr. Creath. The images included the signature spots of cosmic rays; they had not been removed from the images.

In subsequent analyses, however, I created two sets of images, one for each frequency:

1. A set containing the pure biophoton light patterns (with the cosmic rays removed)

2. A set containing only the cosmic rays (with the biophoton and background light removed)

Since I recalled that angels and the Divine Light were supposedly of a very high frequency, I decided one day to explore what might happen if I performed FFTs on the cosmic ray images. Since I assumed that the cosmic rays were random, I assumed the FFTs would reveal smooth, random-like images as well.

I displayed the cosmic ray images and their respective FFT analyses during the presence-of-spirit periods and their respective pre- and post-baselines for all three investigations (the post images look like the pre images).

What transpired can best be thought of as a eureka moment (sometimes called an oh-my-God moment).

I discovered that the FFT analyses on the cosmic ray images, when conducted on the spirit-of-presence periods, revealed the same kinds of ripple-like wave patterns that had been seen for the combined biophoton plus cosmic ray images. The predicted, relatively nondescript, essentially random FFT patterns were observed only in the pre- and post-periods.

Think about this.

The apparent presence of angels and Divine Light created discernable and replicated wavelike structures in FFT analyses of superhigh frequency cosmic ray patterns. In fact, the FFT patterns were more robust for the cosmic rays by themselves than they were for the biophoton images by themselves! To put this in perspective, scientists have always assumed that due to their random patterns and high frequencies, cosmic rays could not be influenced by any outside force, least of all higher spiritual beings.

In science it is often the case that one investigator's noise can be another investigator's signal. I had always thought of the recorded cosmic ray bursts as a nuisance, measurement noise that came along with the supersensitivity of the camera. Who would have guessed that what once was a nuisance might become a valued gift?

A Window of Opportunity Opening?

In the future we will be conducting another series of investigations using the biophoton imaging system. This will include examining the effects of different angelic beings, who purportedly are interested in collaborating with us, and comparing them with those of different deceased beings.

We are asking the question: if angels display a higher frequency of light, and also are more "powerful"—as many have claimed over the centuries—will we find that angels, as a general rule, have more robust effects, especially on cosmic rays, in creating discernable patterns revealed by FFT analyses than typical deceased spirits?

Also, by employing blind experimenters trained to run the camera system, can we rule out the possible role that the experimenter's intention may be playing in producing what we have been hypothesizing may be a presence-of-spirit effect? How much of what we are observing is us versus them?

The reason for sharing these preliminary proof-of-concept observations is to illustrate how it is becoming possible, in principle, to bring higher spiritual energy research into the laboratory. If spirit guides and angels are real, it is high time for them to reveal themselves. They need to show us in a convincing fashion that they cannot only be measured photographically, but that they can be documented to have significant effects on physical life as we know it. Their purported commitment to this research is what I call the Sacred Promise.

Curiously, after completing the preliminary investigations reported thus far, the camera broke. Something happened to the CCD chip; it was as if had been exposed to too much light. We had to send the camera back to the manufacturer to get a major repair. The whole CCD chip had to be replaced at a cost of approximately $10,000. It took a few months for the repair to be completed. Was this a coincidence, or something more? The timing was certainly odd.

I will briefly describe three additional replication investigations that speak to the continued promise of this research. The findings from the second investigation were presented at the 2009 research meetings for the Society for Scientific Exploration; the findings including the third investigation were presented at the 2010 Toward a Science of Consciousness research meetings.

Comparing Angel Sophia's Spirit Effects with Susy's

In this investigation Mark was the experimenter, and he compared the background light levels (averaging the tiny light levels in the visible spectrum) with FFT analyses of the patterns of cosmic rays. He again recorded predata, presence-of-spirit, and postdata collection periods.

On half of the presence-of-spirit trials, Mark invited Angel Sophia into the chamber. On the other half of the presence-of-spirit trials, Mark invited Susy Smith. Since no human subjects (in the physical) were involved, it was not necessary to seek human subjects' protection approval for this research (or for the previous investigations).

Mark ran a total of six sets of Sophia pre-, presence, and post-trials, and six sets of Susy pre-, presence, and post-trials. It took a few weeks to collect this data. When I analyzed the data and compared the FFT and the background light findings, we were amazed.

For the Sophia trials, we clearly replicated the cosmic FFT ripple effects seen earlier. The cosmic FFT ripple effects were statistically significant.

For the Susy trials, no consistent cosmic FFT ripple effects were observed. In a word, Susy did not produce what Sophia did in terms of cosmic ray patterns. Only the alleged Sophia showed effects on these higher light frequencies.

However, for the background visible frequency light analyses in the Susy trials (removing the cosmic rays before doing the analyses), there was a significant increase in the average amount of background light detected. Here we were not performing FFTs on patterns of sporadic cosmic rays; we were calculating averages of background light (to see if there was increased overall visible frequency light in the chamber).

Though the magnitude of the effect was small, it was statistically reliable. The increase in average background light was observed on all six Susy trials.

In the Sophia trials, no consistant increases in average background light were observed. In a word, Sophia did not produce what Susy did in terms of average background frequencies of light (which would be visible to the human eye if their intensity were stronger). Only Susy showed effects on these lower light frequencies. It appeared that only Angel Sophia could affect the higher frequencies of light, or the cosmic rays, and only Susy affected the lower biophoton frequencies.

In the next chapter we will return to the possibility of recording individual photons of light dynamically in real time to detect the presence spirit.

Replicating and Extending the Light of God Cosmic FFT Effects

As to whether the Light of God is potentially a real effect, Mark and I decided to replicate the initial single trial observation, only this time Mark repeated the Light of God trial a total of six times.

In addition, we included six tests where there were no presence-of-spirit intention trials (i.e., blank control trials) as well as six where Mark simply "meditated on the chamber" (but did not invite the Light of God). The question we asked was whether focusing his mind on the chamber was enough to produce the cosmic FFT ripples.

The experimental design was fairly complex: it included a combination of randomization and counterbalancing of the different kinds of trials and conditions. The sessions were run on Saturdays (at times when the lab was quiet, so no other experiments were being run simultaneously). It took a few months to collect this data.

To our utter amazement, not only did the Light of God trials produce significant increases in the cosmic FFT ripple effect, but the two control sets of trials—blank conditions and simple meditation conditions—produced no observable cosmic FFT ripple effects compared to their respective pre image trials.

These findings suggest that there is more to the cosmic ray FFT effects than merely meditating on the chamber or inviting deceased spirits into the chamber. This proof-of-concept exploratory investigation encourages us to design systematic studies using outside subjects approved by the university's IRB committee.

Can Cosmic Ray FFT Effects Occur at a Distance?

You will recall that the camera and light-tight chamber sit in a room that is separate from the room that contains the computer that controls the camera. The experimenter (in these studies, Mark) sat in the adjoining computer room. But, we wondered, can spirit guides, angels, and the Light of God be invited into the chamber by people who are miles away from the laboratory?

In the summer of 2008 at the annual meeting of the Society for Scientific Exploration, I presented the results of a series of six IRB-reviewed experiments involving the effects of distant group intentionality. This involved hundreds of people spread out as groups at varying distances, from hundreds to thousands of miles from Tucson (there was a total of six different experiments), and directing their conscious intentions, as a group, to increase the rate of seed germination in our laboratory.

In the fall of 2009, I was invited to give a keynote address at the International Healing Touch Association meetings. The president, Susan Kagel, a gifted healing touch practitioner and senior medical nurse at Canyon Ranch, asked me if I would be willing to lead a group intention investigation in the middle of my presentation.

I proposed that we try asking the 245 attendees to send their intentions to the CCD camera and chamber (which was approximately fifteen miles from the hotel) for a fifteen-minute period.

She proposed that the healers be encouraged to invite their angels and the Divine to participate. I thought this was a great idea since the cosmic ray FFT effects seem to appear, based on our preliminary findings, only when higher frequency beings were invited.

Mark collected three fifteen-minute trials of pre-, spiritual energy healing, and post- CCD images, and we subsequently analyzed the cosmic ray patterns using the ImageJ FFT image-processing software. The results were striking. At the time of this chapter revision, these results showed the largest cosmic ray FFT ripple effect we had observed to date.

Does this investigation using supersensitive CCD cameras prove definitively that angels and/or Divine Light exist? That they, or their effects, can be measured as increased organization of cosmic rays as revealed by state-of-the-art FFT analyses?

The answer is, of course, not yet. However, this proof-of-concept exploratory investigation points to the possibility that science can address these kinds of questions in a systematic, responsible, and creative manner.

14

THE HOLY GRAIL OF COMMUNICATING WITH SPIRIT

Now what I propose to do is furnish psychic investigators with an apparatus which will give a scientific aspect to their work. This apparatus, let me explain, is in the nature of a valve, so to speak. That is to say, the slightest conceivable effort is made to exert many times its initial power for indicative purposes. It is similar to a modern powerhouse, where man, with his relatively puny one-eighth horsepower, turns a valve which starts a 50,000-horsepower steam turbine.

—Thomas Edison

If Spirit is real and wishes to form a sacred partnership with us for our individual and collective health and evolution, we clearly need a more reliable and trustworthy way of knowing this.

Let's imagine for the moment that Spirit is here. Let's imagine for the moment that the spirits we are dealing with are benevolent, caring, intelligent, and wise. Let's imagine for the moment that we can actually hear them, and what they have to offer is well worth our attention and will make an important difference in our individual and collective lives.

The challenge is, right now we are limited to receiving such information from a small group of psychic individuals who claim to receive such information intuitively—i.e., in their minds. Unfortunately, although some of these individuals can be highly accurate at times, they are by no means infallible. Ultimately, gifted intuitives are human beings, just like us, and they make mistakes nonpsychically as well as psychically, which only clouds the issue. (Others might say that wisdom, unlike knowledge, requires a broader interpretation, thus the ambiguity of some messages, but this a separate issue.)

For example, our ability to hear and accurately remember a routine phone conversation is limited at best, as anyone knows who has tried to recount a phone conversation verbatim, or any conversation for that matter, by memory alone. When we factor in that intuitives are typically only catching snippets of information—sporadic thoughts and feelings—and then are attempting to convey and interpret them on the fly, we can better understand why their accuracy might be decreased accordingly.

Yet, the human capacity for intuition can actually be quite remarkable, especially in gifted people.

For example, as I reported in *The Energy Healing Experiments*, we conducted a double-blind experiment on congestive heart patients and matched controls, testing the ability of a group of medical intuitives to make distant intuitive diagnoses, and to my surprise we obtained statistically significant effects. However, this did not mean that the information was accurate enough to be used as a replacement for diagnostic instruments.

The majority of the medical intuitives who participated in the research claimed that they could not perform this task by themselves. They insisted that the medical diagnostic information they received was provided by their guides.

Curiously, the intuitive with the highest accuracy score happened to be a relatively uneducated grandmother on the East Coast who claimed that she received some medical diagnostic information from Edgar Cayce, whom she called Eddie, and some from specific angels. Given her high performance in the research, should we give some credence to her perceived sources of the information?

Our double-blind medical intuition experiment did not attempt to validate the intuitives' spirit guides explanation for their accuracy. Our IRB-approved university experiment was designed to determine if medical intuitives could make accurate medical diagnoses long distance. The experiment was not designed to discover how they did it. Nonetheless, we can wonder whether their diagnostic accuracy might be enhanced if they could receive the information via some sort of digital technology that provided it in typed or spoken form.

Using emerging technology to detect the presence of spirit is the first step toward addressing what might be called the Holy Grail of the Sacred Promise. I sometimes think of this as the future evolution of the cell phone into the soul phone, or what Rhonda refers to as the future evolution of the typewriter into the spirit writer.

There are numerous claims circulating on the web of such electronic communication technologies. Some claims come from ghost hunters and the emerging ghost-hunting industry. These paranormal investigators attempt to measure air temperatures, magnetic fields, infrared light, and other sources of energy using relatively simple electronic devices such as the trifield meter. Others come from people excited about the possibility of electronic voice phenomena (EVP) using digital audio and video recorders and even computers.

Most of these kinds of claims are so amateurish and naive as to be laughable. Please understand that I mean no offense here; I happen to know a number of the pioneers in EVP explorations, for example, and on the whole they are kind, genuine, and caring people. However, the majority of these spirit seekers—be they ghost hunters or EVP explorers—are laypeople. They have no formal training in technology, nor do they use the scientific method, and their efforts suffer accordingly.

One notable exception is the Windbridge Institute. It has recently embarked on systematic research to detect the presence of spirit with the aid of technology, and the principal investigators come from both the scientific and technology communities. I know of their abilities and backgrounds firsthand; the head scientist, Dr. Beischel, did her postdoctoral training in my laboratory, and their primary technology person, Mark Boccuzzi, has worked for me as a research assistant as well.

As revealed in the previous chapter, my colleagues and I are conducting exploratory investigations in this area, focusing on the nature of light using state-of-the-art photon-detection recording systems. My interest in the quantum and field nature of light, its measurement by interferometers and photomultiplier tubes, and the application of such technologies to spirit detection began when I was a professor at Yale.

It is curious that as the public's use of the awesome capabilities of wireless communication grows, as does its fascination with matters of Spirit and a larger spiritual reality, exciting new technology is emerging that promises to integrate science and spirituality once and for all.

One new technology is the silicon photomultiplier system with its great potential as a spirit detection and communication device. It fulfills Thomas Edison's goal that "the slightest conceivable effort is made to exert many times its initial power for indicative purposes."

Realizing the great promise of this technology, I decided to work directly with the company that manufactures it. At the time I wrote this chapter, I had conducted a series of proof-of-concept investigations myself and was just bringing the technology into my university laboratory. The findings were published in the May 2010 issue of the peer-reviewed *EXPLORE: The Journal of Science and Healing*.

As you will soon see, not only did I explore how Susy Smith and Marcia Eklund could influence the silicon photomultiplier system, but I also investigated whether Angel Sophia and Harry Houdini (yes, you read correctly) could do the same.

But first, let's briefly review the nature of this technology.

The PCDMini's Technology Breakthrough

As Dr. Creath and I summarized in a 2007 paper published in the *Journal of Scientific Exploration* (available on the web), photomultiplier tubes have been used for decades to detect single photons of light. The problem has been that these glass tube systems are very sensitive to surrounding magnetic fields; also, they require high voltages to operate, and they can be broken easily.

However, over the past few years a new sensor technology has been developed, called the silicon photomultiplier, that corrects all of these problems. The devices are relatively insensitive to magnetic fields, they require minimal voltages, and they are much less breakable.

The basic devices—called PCDMinis—are stable, small, and can even be mass-produced. They are extraordinarily sensitive, and they have low "dark counts"—meaning that they generate minimum background noise, which can interfere with the measurements. Another advantage is that they are relatively affordable. In the fall of 2009, a single PCDMini cost approximately $2,500. I reasoned that if a silicon photomultiplier system could be used to reliably detect spirit, it would be priceless.

The dimensions of the sensor and its electronics are approximately one and one-half inches cubed. The sensor itself, which is less than a millimeter in diameter, sits inside the small center ring in the front of the device. The device is cooled to −21 degrees Celsius. Three layers of circuit boards control the sensor system and connect it to a USB port of a computer as well as to a digital oscilloscope.

I initially borrowed, and ultimately purchased, a single PCDMini system to determine if it was sensitive enough to replicate and extend what we were observing with our $30,000 CCD camera. The PCD-Mini has one great advantage over the CCD camera in that it can detect the presence of individual photons occuring in real time. Instead of having to make a recording of fifteen to thirty minutes to visualize the super-low-intensity light emitted by plants—or the even lower intensity light purportedly generated by spirits—the PCDMini can detect changes in photon activity within milliseconds and even picoseconds (billionths of a second).

Of course, an individual PCDMini cannot create a two-dimensional picture like the CCD camera, which has 262,144 pixels. The PCDMini is the equivalent of only a single pixel. However, what the PCDMini loses in two-dimensional imaging, it makes up for in sensitivity and speed of detection.

Remember, the telegraph and its Morse code were binary, which is the equivalent of just a single pixel. So, I reasoned that if Spirit could learn to activate the PCDMini sensor and create blips or bursts on a computer screen, then in principle we could eventually create an electronic keyboard, where each letter had its own sensor (or small array of sensors). Thus, the binary telegraph would be transformed into electronic text. In other words, it would be a Morse code for communicating with spirits so as to get accurate information and guidance. This is what makes all of this experimentation applicable to our concerns.

Three Proof-of-Concept PCDMini Investigations

At the time I wrote this chapter, I had completed three proof-of-concept sets of investigations using the PCDMini system. I first mounted the sensor and its electronics in a box-within-a-box arrangement (plus a small cover), creating a light-tight environment for the sensor. I recorded temperatures inside and outside the boxes to be sure that whatever minor room temperature fluctuations occurred were not associated with changes in the background noise, or "dark counts." In other words, I was testing and validating the company's claim for myself. I was the experimenter.

After the first set of investigations (described below), I built a triple-box enclosure, a box within a box within a box, which created an even greater light-tight environment for the sensor. I initially tested the system in a small laboratory in my home, as I wanted to have ready access to the equipment and data. It displayed the real-time detection of photons on a relatively large flat-screen TV monitor, in addition to my laptop computer screen. The idea was that spirits could have "free play" periods where they learned how to activate the sensor inside the boxes and create measureable effects while watching the screen.

Yes, you read this correctly. Though I do not know whether this was actually the case, I was told by four separate mediums that the spirits were able to influence the sensor inside the light-tight boxes while they hovered, so to speak, above and outside the boxes. Hence, they could supposedly watch the TV monitor as they attempted to influence the sensor.

In point of fact, I made no assumptions as to how Spirit generated photons that the tiny sensor could actually detect (or if it manipulated the sensor in some other way).

I used the basic real-time display and recording software provided with the PCDMini, called the SensL Integrated Environment. The software displays sums of photons, which can vary in size from ten counts in approximately one tenth of a second to more than ten million counts in a tenth of a second. I reasoned that, for both Spirit and me, learning to use the sensor might be like learning to ride a bicycle or play a musical instrument—it would take time and practice.

In the baseline trials, I have observed as low as two tiny photon sums (ten to twenty-five photon counts) in ten minutes (or zero to one photon sums in five minutes). The average number of photon sums was closer to five per five minutes in the double-box and three per five minutes in the triple-box. In presence-of-spirit trials, I have witnessed as many as twenty-four photon sums in a ten-minute period (eight to twelve photon sums per five minutes).

In light of this information, we can now discuss the research. Again, these investigations were to explore if the PCDMini's ability to detect photon sums in real time could eventually be used as a kind of spirit typewriter.

In the first set of investigations, I ran a total of three separate presence-of-spirit investigations, each with matching baseline control trials. In the scientific paper published in *EXPLORE*, I refer to the presence of spirit trials as Spirit Intention, or SI, trials because this was the alleged intention: that Spirit would show up, as invited, and intend to increase the photon counts as detected by the PCDMini, displayed to us as increased numbers of photon sums. The SI trials were compared with baseline trials where Spirit is instructed to stay away from and not influence the sensor.

The first and third investigations consisted of five SI and five baseline trials; the second investigation had ten SI and ten baseline trials. The length of each trial in these three investigations was three hundred seconds (five minutes). In other words, there was a total of twenty SI trials and twenty baseline trials.

The primary spirits invited were Susy Smith and Marcia Eklund. Since I am not a medium, my only gauge of whether they showed up was the positive test results as measured by the PCDMini. The mediums who have read both Susy and Marcia claimed that they were showing up as requested.

It is important to keep in mind that, if positive results were obtained, they could be attributed to my mind itself. To potentially address this fundamental concern, as part of the second investigation, I added ten trials where I, the experimenter, attempted to use my mind to increase the photon counts, with ten matching baseline controls.

Also, I included a final set of twenty trials, all of which were baseline controls as a second set of controls.

The analysis revealed that the Spirit Intention trials had approximately 7.5 photon sums per five minutes compared to 5 photon sums per five minutes for their matching baseline controls.

When photon sums per five minutes are converted into actual photons detected, the values are approximately 150 photons for the Spirit Intention trials compared to 100 photons for the baseline controls.

Not only was this difference statistically significant, but each of the three separate investigations was individually statistically significant.

Point of clarification: Statistical significance means that the observed patterns were reliable and repeated, not that they were necessarily large in terms of magnitude. Though a difference of 2.5 photon sums per five minutes on the average might seem small, the percent magnitude increase associated with the alleged Spirit Intention comes out to 150 percent (7.5 divided by 5 times 100) which happens to match the approximate numbers of photons counted.

When I, the experimenter, tried to produce this effect by myself, I failed miserably. In fact, the experimenter intention trials were actually lower (average of 4.5 photon sums) than their matching baseline controls (5 photon sums), though this effect was not statistically significant.

The no-intention baseline controls mirrored my failed attempts to increase the photon counts.

Could these SI effects be real? Would they generalize to other beings in Spirit? In other words, could any spirit learn to do this, or were Susy and Marcia special cases because they had a strong motivation to connect with their loved ones in the physical, and they had done so for years?

Since I was engaged in exploratory proof-of-concept investigations, I decided to conduct a second set of investigations, and this time I invited two different alleged beings in Spirit, each by itself.

The first was the alleged Angel Sophia, who appeared in the previous chapters. The second was a person who appeared to be interested in presence-of-spirit research, from historical accounts of his time in our physical world. Amazingly (some might say miraculously), the person is Harry Houdini. There are numerous reasons why I considered inviting HHH to participate (HHH is the shorthand I use for the "Hypothesized Harry Houdini"). Discussing these reasons in any depth would be a distraction from the main point of this chapter.

What I can divulge briefly is a particularly convincing private reading I convened a few years ago, where Susy Smith was invited to bring a secret gentleman to the reading, and it was HHH. The medium conducting the experiment, whom I will call Roger, was very confused. Roger said things like (1) "now I see him, now I don't," (2) he could "hold his breath for a very long period of time," (3) he was connected to New York, and (4) his name was not his real name. Because I knew through multiple private readings with Roger that he was an ineffective mind reader (to put it mildly), I seriously entertained that the information the medium was obtaining that clearly fit HHH was probably coming from Houdini (and not from me).

Houdini had been a friend of Sir Arthur Conan Doyle. Where Doyle was a firm believer in life after death, Houdini remained a great skeptic. Houdini became known not only as the world's greatest escape artist but also as the debunker of his day, single-handedly exposing numerous fraudulent mediums.

In these exploratory investigations, I collected four alleged Sophia intention trials with ten baselines preceding and following the SI trials, and eight HHH intention trials with four baselines interspersed. The trial lengths were again three hundred seconds (five minutes).

The findings were clear-cut and statistically significant. Both Sophia and Harry showed increased photon sums compared with their respective baselines. Moreover, their percentage increases in photon sums were approximately 220 percent, compared to 150 percent in the first set of investigations obtained for Susy and Marcia (who supposedly worked as a team).

Could this really be Sophia and HHH? Clearly future research would be needed to determine this, one way or the other.

Meanwhile, I began to wonder if the PCDMini could be used as a binary yes/no device by a skilled Spirit operator. I designed the third proof-of-concept investigation. In my mind, I invited HHH to serve as the SI. The investigation compared the baseline (B), yes (Y), and no (N) response trials. I did not instruct him as to what a yes or no response might look like—I left this up to him. I would then examine the data closely, and see if a reliable difference in burst patterns was observed.

At the onset of each SI trial, instructions in the form of signs for each of the values—baseline/control, yes, no—were presented on the TV monitor for HHH to attempt to make yes or no photon bursts, or to stay away and not influence the counter on the control sign.

Each segment of the four trials was three hundred seconds long. The order of trials was BYN BNY BYN BNY. Whereas the B trials, or the baseline control, always preceded the Y and N trials, the order of the Y and N trials was counterbalanced.

What I requested was that HHH attempt to make one kind of burst for yes and another kind for no. I left the kind of photo sums to HHH.

After I conducted the trials and began analyzing the data, I noticed something peculiar. What appeared to distinguish yeses from noes involved the increased number of photon sums in the first 150 seconds of the Yes trials and the second 150 seconds of the No trials, or the reverse of the previous order. It appears that HHH was adding a further signature to note his presence.

The SI yes effect is obvious to the naked eye; the analysis of the numbers is statistically significant. The percent increase here was approximately 275.

Clearly, something was happening in this investigation. This pattern of results—more yes photon sums appearing in the first 150 seconds of the yes trials, and more no photon sums appearing the second 150 seconds of the no trials—was certainly not part of my consciousness as I served as the experimenter.

In other words, this pattern was not what I envisioned nor expected. We can only wonder, was it really a reflection of Harry Houdini's consciousness?

In light of these findings, the company has loaned me a second PCDMini and a device called an HRMTime module that can measure the time between individual photons to billionths of a second using up to four PCDMini's simultaneously. The timing technology not only increases the resolution of the measurement process but it allows for the simultaneous recording of trains of photons. I designed a software system, implemented by the SensL Corporation, that makes it possible to display in real time two PCDMinis simultaneously and count the number of photons detected per second. I have constructed two identical triple-box enclosures—one will be a yes box and the other a no box.

Will spirits learn to use this system and demonstrate that they can selectively activate the yes or no boxes? If so, the demonstration could be the foundation for developing a technology for creating a real-time spirit phone.

We won't know until they try.

If We Build It, Will They Come?

The skeptic will be quick to point out that all I have done is uncover some potentially promising observations linking my consciousness with the silicon photomultipliers. Remember, in every investigation conducted so far, I invited Spirit into the box—my consciousness was involved, directly or indirectly. But remember that when I tried to create an effect with my mind only, it failed.

Assuming that there are no experimenter errors here, no unanticipated electromagnetic artifacts, and no cheating (and these are well-justified assumptions), the fact remains that we have not established definitively that the system is detecting the presence of Spirit—be it Susy's, Marcia's, Sophia's, Harry's, or whomever's—independent of the experimenter's consciousness.

Future systematic research can include specific experiments that are run by ardent skeptics; the fact is that cell phones work whether we believe in them or not, and the same should apply to future soul phones. Future research can also include having the PCDMinis run autonomously by the computers themselves—in the absence of any human experimenters; specific spirits will be invited to participate in SI and control trials as indicated on the TV monitor screen that is never observed by a person in the physical. At the time this chapter was being copyedited, we had completed a series of computer-automated experiments, and the results continued to be positive. A paper titled "Possible application of silicon photomultiplier technology to detect the presence of spirit and intention," describing these findings, was accepted for publication in *EXPLORE*.

I trust you realize—given what you have read in this book—that if I were asked the question, "Are you trying to prove that Spirit exists and that it can learn to use this equipment?" my response would be: "Absolutely not. What I am trying to do is give Spirit—if (1) it exists and (2) it can learn to use this equipment—the opportunity to prove it itself."

My job as a scientist is to provide optimally sensitive and responsible conditions for data to be discovered. I am merely the experimenter. The rest is up to them.

In the movie *Field of Dreams*, there is the famous line, "If you build it, they will come." Metaphorically, my job is to build it. If the premise of this book is correct, the Sacred Promise is not only that they will come, but they will do so with rings on their fingers and bells on their toes.

A Proof-of-Concept Anomaly

I end this chapter with an amusing anecdote that brings us back to the beginning of the book. You will recall how Susy Smith allegedly showed up at the home of a medium I called Joan, and that the first personal test I did to determine whether Susy was potentially watching over me was to view the movie *Field of Dreams* while eating Chinese food and reclining in bed as a random event for her to report. You will recall that *Field of Dreams* is about a man who builds a baseball field in his cornfield, and famous deceased baseball players—as well as his deceased father—come and use it, gratefully and enthusiastically.

The anecdote—which I prefer to call a proof-of-concept anomaly—happened during the period of time that I began to wonder whether Harry Houdini was participating in the second exploratory investigation.

It was a Wednesday, and I was having a private meeting with Jerry Cohen, CEO of Canyon Ranch, about the current progress of the Voyager Program. I left the PCDMini system at home with the TV monitor running and informed them, our friendly spirit collaborators, that this was a "free play period"—when they could practice using the system in any way they wanted. These playtime periods were made available at various times. Though I videotaped the TV monitor using time-lapse recording equipment, I did not analyze the data since it was completely uncontrolled.

At this meeting, I had prepared a short video plus a PowerPoint presentation about the possible use of the PCDMini system, and I briefly mentioned the hypothesized Harry Houdini possibility. It turned out that Jerry was a Houdini fan, and he asked the following question, which I paraphrase:

"If Harry is that good, can he hit a home run? Can he make larger spikes than just twenty-five- or thirty-unit photon sums?"

I told Jerry that every now and again, a larger photon sum would be observed—fifty or even seventy-five units large—but I had never asked them to try and make larger ones, or smaller ones, for that matter.

As I drove home, I remembered the comment that a secret sitter once made about Mary Occhino. He said, "Mary not only hit the ball out of the park. She hit it out of New York City."

I wondered, could Harry hit a home run, on command, and could he hit it out of the park, if not the city of Tucson? I had a meeting at my home scheduled with a postdoctoral fellow, Dr. Jolie Haun, and I was ten minutes late. I raced in and quickly looked at the TV monitor in my study. To my amazement, the sporadic photon sums on the screen were tiny, less than one-fifth their normal size!

Typically the graphs on the screen showed twenty-five-unit-high photon sums, which reached the full height of the Y-axis. Why did the photon sums now look so tiny? Had the software automatically adjusted the Y-axis scale because a larger photon sum had appeared? (Which is typically what would happen.)

When I looked at the Y-axis, I noticed that the scale was not reading 0 to 25; it was reading 0 to 175!

In other words, the new 0–175 range implied that at least one photon sum had occurred that was possibly as large as 175 photons. Remember, virtually all the photon sums in the pitch-black chamber had been only 25 units high or less. A burst of 175 photons would be 700 percent greater than a typical burst of 25 photons!

Since Dr. Haun was not aware I was conducting this personal research, and since our meeting was not about life after death or spirits, I simply noted the time. Later I would review the digital time-lapse photography and see when the superlarge photon sum occurred—presuming it was a photon sum and not an artifact.

After our meeting, I replayed the time-lapse video and discovered that a 173-unit photon sum had occurred just around the time that I was driving into my garage.

I wondered, could this be an intentional response by HHH? If something happens once, it could be a chance event, an artifact, something unimportant. But if this phenomenon was real, somehow attached to HHH, he should theoretically be able to duplicate it.

I decided to restart the free-play period and watch what unfolded. The time-lapse camera was taking snapshots of the screens. As I was watching the screen, I asked Houdini, in my head, if he could make a large photon sum.

What happened next I witnessed with my own two eyes.

In watching hundreds of hours of outputs, I rarely saw photon sums above 50–75. I was now seeing a 173-unit photon, and it occurred shortly after I had asked Houdini if he could make a big one!

Was this just a coincidence? Could it be something more? I did not know. I had to give a lecture at Canyon Ranch that evening, so I left the system running in playtime. When I returned around 10 PM, I noticed that the Y-axis was back to normal: 25 units.

I was really tired, but I wanted to watch a little more. Around 10:15, I decided to ask Houdini if he could make another "big one." This time I invited Rhonda to be present.

To my and Rhonda's utter amazement, another huge 173-unit photon sum appeared on the screen.

You will recall in the chapter where I discussed the spontaneously double-blind spiritual healings I received—both the first and second times that my symptoms subsided synchronously with my receiving a secret spiritual healing—that I thought the timing might be coincidental. However, when it happened a third time, it was the charm. At that point it seemed prudent to seriously ponder whether there was something nonrandom and real occurring.

One big photon sum, coincidently after Jerry asked about home runs and I wondered about spirits playing baseball and a medium hitting the ball out of New York City, could have been chance.

A second big photon sum, occurring within a minute of me asking in my head if Houdini, if he was here, could hit another home run could have been a chance event, too.

However, here was a third big photon, again occurring within a minute of my asking in my head, in Rhonda's presence, whether Hypothesized Harry Houdini could hit the ball out of the park. Was this chance, too, or was this the charm? In fact, had HHH metaphorically hit the ball out of the city of Tucson?

As Susy Smith reminds us, sometimes it is "too coincidental to be accidental."

It is one thing to read about this anomalous anecdote; it is another to actually experience it. The truth is that there is no substitute for being there.

Consider this example.

It is one thing for me to describe the rings of Saturn.

It is another for you to see a picture of Saturn's rings.

However, neither of them holds a candle to actually being outside, in the dark, looking through a contemporary, computer-controlled optical telescope and experiencing the glorious rings of Saturn with your own eyes. There is no substitute for the real experience.

For the record, at present I do not know if the measurement of individual photons of light, under the proper controlled conditions, will evolve into a practical and accurate technology for detecting spirits and communicating with them. Like the Windbridge Institute, we are exploring other possible measurement candidates, including the modulation by Spirit of super-low-level magnetic fields and even the modulation of frequencies of radio waves in an electromagnetically shielded environment.

GARY E. SCHWARTZ, PHD

Preparing for the Ride of Our Lives— a Wright Brothers Moment

I have come to believe that we are close to experiencing a Wright brothers moment with regard to the technology of detecting Spirit.

For thousands of years humankind dreamed of flying. But it was not until that fateful day a little more than a hundred years ago at Kitty Hawk, when the Wright brothers' motor-powered airplane flew for all of twelve seconds (the duration of its first flight), that we discovered that powered flight was possible.

The Wright brothers' plane took flight a total of four times that day. The next time they tried it, the replication failed—apparently their primitive flying machine required the assistance of wind, and the air had become too calm.

Though it would take a few decades for regular commercial flights to become an almost-everyperson reality, humankind's vision of what was possible was forever changed.

If this chapter, and the previous one, is any indication, we may be approaching a Wright brothers moment as we awaken to the reality of using technology to create sacred partnerships.

We all remember Neil Armstrong, the first man to walk on the moon. I have begun to wonder, who will be the first spirit to talk to us from the other side?

Will it be a relatively unknown person like Susy Smith? Will it be the great showman and skeptic Harry Houdini? Could it be the loving and gracious Princess Diana? Will it be the distinguished scientist Albert Einstein? Or will it possibly be the controversial superstar who wrote the inspiring song "We Are the World," Michael Jackson?

If the Holy Grail of the Sacred Promise is eventually realized, I would not be surprised if we ultimately hear from all of them, as well as millions, if not billions, of others.

Of course, if and when that day ever comes, we will have the even greater challenge of discerning whom we should listen to. If the resulting "Spiritnet" is anything like the internet, we will have our hands and minds full—literally.

That is why developing our consciousness, or intuitive faculties, will always be of the utmost importance in our ever-widening exchange with Spirit. Or, one might say, it is the ultimate litmus test. This is evident when reading so-called channeled material: some of it is very informative, but a lot of it is questionable, if not completely erroneous.

How do we separate the wheat from the chaff? Through our own intuitive discrimination, which is confirmed by external evidence. I would like to conclude this book with the story of a woman taking that journey as inspiration for us all, because the real value of this contact is how it elevates our consciousness and improves our world.

15

LEARNING TO CONNECT WITH SPIRIT

The love of learning rules the world.

—Phi Kappa Phi motto

Can people learn to connect with Spirit?

Can people learn to use this information in ways that provide practical guidance in their lives?

Can people learn to apply the scientific method for improving their ability to discern the difference between what is real, what is possible, and what is imaginary?

And if so, how will we conduct formal research on this question?

What you are about to read is the inspiring journey of a person who decided to develop her intuitive skills for connecting with

Spirit. This person has a deep questioning mind and appreciates the value of science. And it was fortuitous—if not synchronistic—that she was hired to work in a laboratory that was actively doing such research. Both she and I happened to be in the right place at the right time, doing the right thing with the right people.

It is an honor for me to provide a brief account of her journey, because I can appreciate her courage in having chosen to share her journey at all, let alone in doing it so publicly because she knows how much it could benefit others.

Her name is Clarissa Siebern, and at the time I wrote this chapter she was in her late thirties and working as the program coordinator of the Laboratory for Advances in Consciousness and Health at the University of Arizona. She had also assisted in certain research projects in the laboratory and is the mother of a boy with physical disabilities who appears to have some natural healing abilities.

It is curious how Clarissa came to my laboratory.

Clarissa was hired by the previous program coordinator, whose responsibilities included administering my Center for Frontier Medicine in Biofield Science, which was funded by the National Center for Complementary and Alternative Medicine, part of the National Institutes of Health. Clarissa served as her assistant. I rarely saw her.

When the center's funding ended and my former program coordinator moved to another department in the university, Clarissa became my assistant and was subsequently promoted to the position of program coordinator when I obtained some private funding to continue the work.

In other words, Clarissa was not hired because she had intuitive leanings. Moreover, I had no idea that she would later decide to obtain personal training as an intuitive and end up becoming an intuitive investigator in the laboratory. Clarissa began her formal intuitive mentoring experiences in 2005, and it has been a surprising and remarkable journey.

You might wonder what it is like to be Clarissa and take such a journey yourself. What is it like to feel an increasing presence of a group of loving and caring guides in your life? Just as yours would be, Clarissa's journey was a matter of building trust while building bridges.

What is it like to experience your guides seemingly showing up in your car, in a research meeting, or in the bathtub, and having them give you guidance, sometimes quite firmly? Clarissa says it's rarely intrusive and often in response to a silent prayer or request.

What is it like to sometimes do tasks triggered by your guides, seemingly out of the blue? This would include purchasing two books about the Imagineers at Disney for your boss and then an hour later learning that he had just spoken with a colleague who had recently met the Imagineers' president.

How do you feel if you are learning how to communicate with Spirit, and someone in the physical you care about—in this case, your boss as well as your friend—is constantly questioning whether you are getting accurate information from your alleged guides?

This situation is pretty unique to Clarissa; she describes it as being under a magnifying glass. I'm not sure that I would be as courageous, patient, and understanding as Clarissa has been under these circumstances.

I will first share four examples that I witnessed that clearly and verifiably illustrate how Clarissa's apparent partnership with her guides led her to make specific (and often surprising) intuitive leaps. Since Clarissa has been engaged in evidence-based intuition training, I have had the privilege to serve as one of the people who helped her verify which of her intuitions are valid. The examples also demonstrate how guides seem to play a role in contributing to and validating synchronistic events.

I will then review two examples of exploratory proof-of-concept investigations, which are part of the Voyager Program and in which she participated as a co-investigator. These investigations illustrate the potential effects of her intuitive training as well as the promise of this work.

How Clarissa Receives Guidance from Her Guides

Over time I came to discover, and then expect, that I would have regular (at least monthly), memorable, guide-related validating experiences with Clarissa. She seemed to have become so adept at communicating with her guides that she would do things that were uncannily timely and surprising, if not jaw-dropping in their accuracy and significance.

We have probably experienced more than a hundred specific, meaningful, and verified events over a four-year period purportedly initiated by her guides. Moreover, the frequency and complexity of these events have increased significantly over this time period as well.

The increase might be explained as Clarissa feeling more comfortable about expressing the information and/or me being more open about receiving it. However, some of it may be because Clarissa is becoming better at listening to, questioning, and communicating with her guides.

I include four noteworthy examples below, one per year, to give you a flavor for the nature of her gifts and the apparent communications that she experiences.

On a Sunday afternoon in 2006, I decided to purchase a new watch. I had not bought a new one in more than twenty years. The purchase was spontaneous. I felt moved to purchase a silver-and-gold Bulova watch with a dark blue dial. I could not understand why its striking blue face appealed to me, since I had previously worn a conservative stainless steel Rolex with a plain white face.

Then on Monday morning, Clarissa called to tell me that she had been moved by her guides to purchase a gift for me on Sunday—

around the same time that I had bought the watch. She insisted upon coming to my house to present me with it. She had never felt the urge before (or since) to spontaneously give me a gift that she had to bring to my house.

The present was a print of the painting *Meditative Rose*, by Salvador Dalí. It depicts a large red rose appearing like the sun against a vibrant blue sky and seemingly floating above golden-colored mountains. I was shocked. The dark blue portion of sky of the painting was the same color as the face of the new watch I had been moved to buy (I had not at that time noticed the possible golden mountains/gold water connection as well).

Clarissa had purchased the print not only at her guides prompting but also because it jogged her memory of some synchronicities I had had in 2005 and earlier in 2006 involving single red roses. In other words, though her memory of past synchronicities made her open to purchasing the painting, it was the urging of her alleged guides that provided the motivation to actually make the purchase.

Though I knew nothing about Salvador Dali's paintings then, I deeply appreciated receiving her gift, especially given the apparent connection of its striking blue background with the blue of my new watch's face. This gift really caught my attention, and I hung the print in the hall outside the master bedroom.

Five days later at a conference held at Duke University where I gave the keynote address, Dr. Larry Dossey, a distinguished author and speaker, prominently displayed the same *Meditative Rose* by Salvador Dali on one of his PowerPoint slides.

In more than thirty years of attending scientific meetings, I had never seen such a painting in anybody's research presentation, and I pondered on the odds of it being the same as a gift I had just received.

I later asked Dr. Dossey why he used the image. He said that he happened upon it on the web a few weeks earlier but just that morning felt the urge to include it in his presentation. He said it had no specific meaning to him or the content of his scientific presentation, other than it was beautiful and made him smile.

The true import of this synchronicity became apparent approximately two hours later. I learned that a woman I had just met—who subsequently became my wife—had been drawing rose-colored suns and that her first name, Rhonda, means "rose" in Greek and her middle name, Rae, means "sun"!

It was one thing for Clarissa to seemingly chance upon the poster at the mall on that Sunday morning. It was another thing for her to feel the urge to purchase it for me. And then she felt the need the next morning to drive from the laboratory to my home to present me with it because she could not wait for me to arrive at the laboratory. All of this emphasized its potential importance to me. (It is possible that Clarissa's alleged guides really wanted me to take notice of Rhonda when she showed up in my life, and if so, I bless them for it.)

Clarissa claims that she increasingly feels encouraged by Spirit to do certain things that turn out to be evidential and meaningful.

The second example happened in the fall of 2007, when Clarissa felt the urge to share with me some words of wisdom that presumably came from her guides. According to Clarissa, they said that one way to think about them was as providing nurturing similar to how a parent assists a child learning to walk. Clarissa said, "A child will fall and get bumps and bruises, but the parent is there for support to make the journey less painful. You are the child and they—the guides—are there to assist."

I found the timing of this spontaneous wisdom from Clarissa's alleged guides quite odd because just that morning, when I awoke, I happened to find a large bruise on my leg and I could not figure out where it might have come from. I had discovered a bruise on my legs upon waking maybe three or four times in my entire life.

Are guides like caring parents? Can they assist us metaphorically in learning how to walk?

I realized that the Universe was potentially trying to get my attention with Clarissa and her guides' help. I was being reminded to sustain a childlike openness and wonder concerning future evidential guidance and to preserve and protect my youthful caring heart.

The third example also involves children.

In the summer of 2008, Rhonda and I went to Hawaii to attend a one-year memorial service for a distinguished Hawaiian psychologist and dear friend, the late Dr. Paul Pearsall. At his memorial service, I met a Hawaiian spiritual leader whom Paul had loved and admired, Frank Kawaikapuokalani "Kuma" Hewett. Paul had written the foreword for this book, which introduced me to Kuma's apt quote: "If you are going to create more light for our world, you must be willing to endure a little heat."

Kuma introduced me to his daughter, who happened to have a teddy bear with a rainbow T-shirt sitting on her lap. I took some pictures because I had been having synchronicities around bears and rainbows on this trip. Only Rhonda, Paul's wife, Celeste, and I knew that this combination of events was occurring.

When we returned to Tucson, Clarissa came to our weekly meeting with a gift that her guides supposedly had instructed her to purchase for me. I could not believe what I was receiving.

Clarissa gave me two Care Bears, one with a heart and one with a rainbow! I subsequently gave the teddy bears to a child as a gift.

How often do you think she (or anyone) has given me a teddy bear as a gift, especially with a rainbow on it? And how often do you think I see a child holding a teddy bear with a rainbow and take a picture of it? The probability is minuscule.

According to Native American mythology, the bear is the symbol of the power of introspection. According to the book that accompanies the *Medicine Cards*, the bear resides in the west, which represents the intuition. On occasions when I have become imbalanced—either valuing the cold reliability of technology more than the warmth and caring of spirituality, or embracing the comfort of logic and reasoning over the spontaneity and novelty of intuition—I have been gently yet firmly realigned by unanticipated synchronicities, including direction from Clarissa's alleged guides.

The final example I have selected also happens to involve Hawaii. I include it because it also relates to John Nelson, the gifted developmental editor of this book.

In the early fall of 2009, my publisher selected John to work on this book, and we had our first conference call one morning. It was at that time that I learned that John happened to live in Hawaii. Save for the publisher and team at Beyond Words, my agent, and Rhonda, no one else whom I was aware of had been told that John was the book's developmental editor and that he lived in Hawaii.

That evening Clarissa, Rhonda, and I attended a meeting. Here is how Clarissa later recorded the event in her unedited notes:

> I was told [by her guides] to bring Hawaiian sweet bread and decorate our Monday night meeting in a Hawaiian theme. There were palm trees and hula music and pineapple. I had no idea why. I learned that, as so many times before, usually Gary would be sharing something or that it related to something. He was unaware of the theme as he shared that he had just discovered that an individual related to the publication of his book either lives or happens to be in Hawaii.

Interestingly enough, while I was ready to charge off and just duplicate the first draft for him to edit, John insisted that he do an overview before we proceeded. I initially resisted this despite our Hawaiian synchronicity (what John coined for us as "Spirit's calling card"). However, I quickly came to appreciate the direction he was suggesting.

After four years of witnessing such events, only a die-hard disbeliever would dismiss the possibility that something real was going on.

The nagging question remains, what is the source of this information? Is Clarissa correct that at least some of it is coming from her guides?

Clarissa's Participation in Self-Science Research

Because of confidentiality issues, I cannot share information about Clarissa's intuitive mentors or their mentoring styles. What I can say is that they are professionals who happen to be trained in the mental health field and that they are responsible and successful people. And yes, to various degrees, they hold some strange beliefs. For example, one of them is convinced that she has guides that have existed since the beginning of time, i.e., they are eternal beings of light and information, and they assist numerous people on this planet.

Since Clarissa works in a research laboratory and because she has a strong inquiring mind, she has actively assisted in the design of exploratory proof-of-concept investigations as well as helped us pretest the viability of conducting future systematic research in the area. She feels that this opportunity to voluntarily test herself has been invaluable to her because it balances her personal side with her professional side. Most intuitives-in-training are not exposed to the active combination of both approaches.

What follows are two investigations that illustrate not only Clarissa's gifts and emerging skills but also the promise of scientists conducting future research in this area.

The first involves the possibility of testing whether spirits can assist humans in influencing the physical world and/or can influence the physical world themselves. In brief, we (Clarissa, other colleagues in the laboratory, and me) conducted an exploratory investigation on ourselves to determine the possible effects of intention on the growth of seeds. Previous systematic collaborative research between Lynne McTaggart, author of *The Intention Experiment*, and my laboratory has explored the possible effects of a group's distant intentionality on the growth of barley seeds. The six experiments were conducted with a double-blind protocol.

In the spirit-assisted exploratory seed growth investigation with Clarissa, we compared (1) Clarissa attempting with her intention alone to influence the growth of seeds, without the assistance of her guides, which we called "the me condition"; (2) her guides attempting to influence the growth of seeds, without the assistance of Clarissa, which we called "the them condition"; and (3) the combination of Clarissa and her guides attempting to influence the growth of seeds, which we called "the us condition".

In this investigation, the seeds being tested were mung beans. We purposely asked the alleged guides what kind of seeds they would prefer to influence, and they supposedly selected mung beans. I later learned that the sprouting of mung beans has been studied extensively by plant scientists and that mung beans are relatively easy to grow and measure. None of us knew this at the time.

Each intention period was twelve minutes in duration, performed three times a day for four consecutive days. The tests were typically performed at Clarissa's home or on lunch breaks. The seeds—forty per condition—were grown under carefully controlled conditions in a separate location. There were numerous nonintention control sets of forty seeds as well. The seed growth and scoring (length of sprouting in millimeters) were performed by Dr. Robert Stek at his home. This was all personal exploratory research as a prelude to possibly bringing it into the university.

The investigation was performed two times. The combined results of the two investigations were encouraging and point to the possibility of conducting future systemic research in this area. The summary graph for Clarissa serving as the experimenter (the seeds were technically the subjects) is shown below. Each bar reflects the average length of the intention-targeted seeds (the me, them, and us conditions) minus their matched, nontarget control seeds.

CHANGE IN MUNG BEAN HEIGHT VERSUS CONTROLS

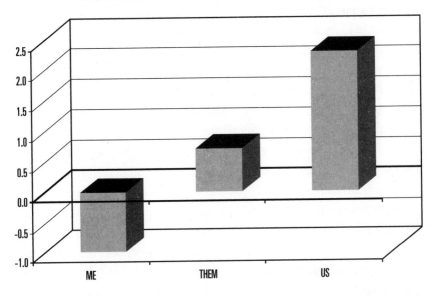

You can see that by herself (the me condition, left bar), Clarissa's intentions were not associated with an increase in mung bean growth compared to matched control seeds. If anything, her mung beans were slightly smaller in size than the matched controls. It is important to note that Clarissa did not believe that she could affect seed growth, especially from a distance, by herself.

You can also see that the results for the alleged guides (the them condition, the middle bar) were associated with a slight increase in seed growth compared to matched control seeds. Clarissa believed that they might be able to have some effect.

However, you can see that it was the combination of Clarissa and her alleged guides (the us condition, the right bar) that was associated with the largest increase in seed growth compared to matched control seeds. This finding made Clarissa happy because it supported her personal predictions. It also suggests a reason why Spirit may be showing up in our lives, that its effect on the world must be mediated through human contact or come through the conduit of conscious human beings.

Do these findings establish that Clarissa's guides, especially with her assistance, can have a measurable effect on the growth of seeds? Even though the findings are statistically significant, there are possible alternative interpretations of the findings.

For example, the question could be raised as to whether the findings simply reflect a belief effect. In other words, is it possible that the observed effects were caused by Clarissa and her beliefs, and that the purported guides, even if they do exist, were not causally involved in the effects observed?

Clarissa is mindful of the fact that these positive findings, which clearly support her beliefs, do not necessarily establish that her beliefs are true. Moreover, she understands that findings obtained in a single, clearly motivated person—i.e., herself—do not establish general application. Without doing future research, we cannot know if such findings will generalize to other people (with their specific beliefs and guides).

However, these observations illustrate the promise of conducting such future research. And for Clarissa personally, they provide her with exciting reinforcement that what she appears to be experiencing can be demonstrated in the real world.

A second illustration involves quantifying what is happening inside a person's brain when he or she communicates with alleged guides, and even while attempting to make mung beans grow.

In the spring of 2009, we had the opportunity to pretest state-of-the-art computerized brain-wave monitoring equipment to track moment-to-moment changes in EEG patterns (1) during Spirit communication versus resting periods and while (2) sending intentions for distant seeds to grow. The equipment was operated by a registered nurse and certified practitioner using Brain State Technology equipment. The EEG equipment is safe, noninvasive, and provides immediate analyses of EEG frequencies in real time. Again, the setting was a private home, but the opportunity afforded by these self-science investigations could potentially set the stage for formal experiments at the university level.

Clarissa was eager to have her brain scanned. The operator placed surface electrodes on the left and right occipital regions of her head (the back of her head). We discovered that during eyes-closed resting periods, her brain showed a prototypic pattern for someone who has developed intuitive abilities: increases in very low-frequency delta (one to four cycles per second) and sub-delta (below one cycle per second) frequencies. In fact, her brain looked surprisingly similar to Dr. Howard Hall's brain, as discussed in chapter 7.

During eyes-closed periods when she was silently connected with her guides—speaking out loud was not feasible because it creates scalp-muscle artifacts in the recordings—her brain showed an increase in very low delta frequencies as well as increases in bursts of high-frequency beta activity (forty to sixty cycles per second), which are typically associated with increases in conscious awareness.

It was also interesting that she showed a greater increase on the right side of her brain when compared to the left. The reason this is surprising is because the left hemisphere is typically associated with verbal, linguistic, and logical processing, whereas the right hemisphere is associated with intuitive, creative, and more visual processes. The EEG patterns were not consistent with the hypothesis that communicating with Spirit is simply a verbal and logical process; it appears to be more intuitive and creative.

Also interesting was what happened when we asked Clarissa to send intentions to distant mung beans by herself versus involving her guides. When attempting to send intentions by herself, her brain showed decreased delta and sub-delta brainwaves as well as decreased high-frequency beta, compared with a resting baseline.

When she invited her guides to help make the beans grow, the EEG pattern showed a relative increase in delta and sub-delta brainwaves as well as a relative increase in high-frequency beta.

In other words, the state of her brain during the joint sending of energy to the beans looked similar to the state of her brain when she was intentionally, silently communicating with her guides.

As a control test, Dr. Stek wanted to see what his brain readings looked like and how they compared with Clarissa's. We observed that his brain looked more normal in the sense that his predominant EEG eyes-closed resting EEG pattern was in the alpha range (eight to twelve cycles per second). When he tried to imagine communicating with his guides—he claims no conscious awareness of such guides, nor has he been instructed on how to be connected with Spirit—his brain showed a slight increase in low-frequency beta activity (thirteen to twenty cycles per second) but very little of anything else.

It would have been valuable if we had comparable EEG brainwave recordings of Clarissa before she began her intuitive mentoring. We can only wonder if we would have discovered that with training and experience, her brain showed an increase in the very low delta frequency and the very high beta frequency patterns over time.

Can the future study of intuitives' brains help us better understand the nature and mechanisms of connecting with Spirit?

Right now this is only a promise; such future science is waiting to be manifested.

Living with Intuition and Awareness

Clarissa feels that her personal intuitive experiences—complemented by her professional laboratory work—have been inspirational and meaningful. Here is how Clarissa summarizes her experiences, followed by her current understanding of them.

> I have had a unique opportunity to not only work around this research but to meet and work alongside many of the respected intuitives that Dr. Schwartz has researched and collaborated with. I can recall that from a very young age I was intuitive at certain levels. However, it became apparent that over time my abilities for precognition, intuition, and other faculties were amplified by this environment. I have had at my disposal the ability to test and verify information received, dreams, and various other forms of intuition within designed investigations, and have even been able to apply intuition in assisting with missing persons.
>
> To some this may appear to be a Candy Land of exposure; however, it is not without disadvantages. The explorations involve challenging and continuously questioning theories and even beliefs. It can be taxing and has tested my limitations. Having had Catholicism as my religious foundation, this has not only challenged my religious beliefs but has also pushed me to take inventory of my moral and spiritual values. This takes a lot of personal work and strength to peel away the layers of myself, and to put aside preconceived ideas and notions, only to rebuild by throwing out or including new beliefs and information. This becomes the power and meaning of integration.
>
> I acknowledge that I have a number of individuals that I can request feedback and counsel from. All are in psychology fields. I have a friend who is a neuropsychologist, my mentor who is an intuitive and counselor, colleagues who are either clinical psychologists or research psychologists. They are a stabilizing force that allows me to experience this work that

has been beneficial to me, while maintaining an understanding regarding the unknown.

Intuition is like baseball. There are good games, and there are days that I want to throw the bat away. There are times when you have a home run or may even hit the ball out of the ballpark. My experience is that information can be received from a deceased individual that comforts a family member, a precognitive dream that is later validated, or even just a simple message.

As in the game, where you have to still run the bases, it is imperative that you are able to deal with the nature of the experiences, whether it is grief, fear, or even joy. I relate to these experiences by utilizing my personal history and how it correlates to my life. As an intuitive and sensitive, I handle the experiences additionally on an energetic level. As I have previously described, the work around challenging your beliefs and integrating them is the most difficult aspect of this development.

Frequently there is no clear understanding of the source of the information. It is like trying to catch a fly ball in a night game with the lights blaring. Some people call it a gut feeling, some call it intuition, some the knowing field, some the Akashic records, and some even feel it is just a coincidence. I feel this is often the best way to receive information. I am able to focus on the content of the information or message and not get hung up on the method of delivery. Knowing who the source is can often be just a distraction to the intent of the message.

Who is the umpire? Is it your conscience, your religion, God, the Universe, the Source, a scientist, or yourself? Are you a pitcher, a catcher, or a coach?

In my experience I have come to discover my own strengths in various areas of intuition. There are some areas that are weaknesses for me as well. Not every player can support the entire team on his or her own.

At the time I wrote this chapter, Clarissa was continuing her intuitive education, along with being a university laboratory program coordinator, mother, and wife. Her understandably skeptical husband was slowly beginning to appreciate that his wife was not strange but that she had a genuine intuitive ability that may have real value in the world.

I hope, and anticipate, that Clarissa will write a book someday that recounts her personal journey of discovery and awakening. Clarissa is a true self-scientist, both in her personal as well as her professional life. If Spirit is real, it appears that Clarissa is getting a lot of assistance, big time. And so might everybody else who sincerely takes that first step by opening to Spirit, charting the synchronicities that appear, and interpreting them in a clearheaded manner. As with Clarissa, more will be given to you as your personal journey to Spirit begins.

AFTERWORD

Spirit Is Calling Us

What's wrong, Mom?
Hmmm . . . You ought to go downstairs and . . .
pray to your guardian angels.
It helps me.

—Anonymous teenage boy

The day before I was planning to write this afterword, an associate, who requested anonymity, shared a piece of advice from one of her children that almost brought me to tears. I will call her Carol, her son Evan, and her ex-husband Eugene.

The timing could not have been more propitious. As I heard Carol tell her story, I was reminded that the Sacred Promise is not just for adults and the earth today, but it is especially important for our children and their children, and hence the future of our species and the planet.

AFTERWORD

To place her son's advice in context, a bit of history is helpful. (I have modified a few insignificant details below to protect the young boy's and his family's identity.)

Evan had been experiencing great emotional challenges over the past year. He had what could be described as visions, as well as sporadic bouts of uncontrollable anger and anxiety. His mother had taken him to both conventional and alternative practitioners, including pediatricians and child psychiatrists, spiritual counselors and energy healers.

Evan was not using recreational drugs; he is an excellent athlete and musician and an all-around good kid. The consensus was that he needed medical intervention to help control his anger and to reduce his stress, and he was placed on appropriate medications that helped suppress his emotions as well as some of his visions. He was grateful for the pharmacologic assistance, but missed some of his positive inner spiritual experiences that also were quelled by the pills.

Sadly, his divorced parents could not be more different in their beliefs or approaches to this challenge, and they were fighting over who had the boy's best interests at heart.

His mother, Carol, is religiously oriented and has a loving faith in a larger spiritual reality; his father, Eugene, is hostile toward spirituality and is convinced that experiences such as communicating with guardian angels are "psychotic."

Evan had been evaluated by Dr. Jackson, who as discussed previously, allegedly grew up experiencing communication with guardian angels. The boy demonstrated striking evidence of having psychic capabilities, including the capacity for accurate remote viewing of objects as well as identification of people in spirit.

Who is Evan to believe: his mother or his father, which doctor?

The boy loves both of his parents, and he needs loving acceptance and approval in return from each of them.

AFTERWORD

I could only imagine what it must be like for each of them: a mother loving her son and believing in a larger spiritual reality; a father loving his son and believing that the physical world is all there is; and a teenage son loving his mother and father and being stuck in the middle of their conflict. My heart went out to the three of them.

Following an upsetting session at a court hearing, Carol broke down in tears. What happened next is summarized below in the email I requested from Carol about her experience, slightly modified to preserve her anonymity:

> As I walked into my son's room to say good night, he was reading his Harry Potter book in his bunk bed. He noticed that I had been crying and he asked, "What's wrong, Mom?"
>
> I said, "Well honey, sometimes things are overwhelming and I am trying to just sort it all out, and I am worried about you."
>
> He looked sideways from his book at me and said. "Hmmm ... You ought to go downstairs and ... pray to your guardian angel. It helps me."
>
> And he went back to reading without missing a beat.

Once again, the doubt resurfaced in me, and the questions reappeared.

Was this child deluding and protecting himself by imaging the equivalent of Santa Claus and the Easter Bunny?

Or was this potentially psychic, if not gifted, child experiencing and doing what each of us should be doing rather than censoring the impulse—calling on our spirit guides with discernment and with the help of the Sacred, especially when we need assistance?

251

AFTERWORD

Do all of us have this opportunity to connect with a larger spiritual reality, and if we do, can we count on the spirits being there to help protect and guide us?

If you take the proof-of-concept exploratory investigations and experiments reported in this book as valid, then the results from future definitive research will not come soon enough for this young boy and his parents. Evan and his parents want to know as soon as possible if this is true, and so do I.

If you have read this far, I suspect you want to know, too.

If Not Now, When? If Not Us, Who?

If Spirit is not only with us, but is calling on us to wake up and see the light, then if ever there was a time for us to get off our collective duff and get to work, it is now.

If you consider the essence, the big picture, of what you have read in this book to be true—recognizing that some experimental approaches may be out there—then there is the serious possibility, if not probability, that the spirits are awaiting our active collaboration with them.

They are here with you and me right now, awaiting our awakening.

Not only that, but they are claiming to be at least as dedicated, trustworthy, genuine, and capable of helping us save the planet as the most highly evolved, responsible, and caring individuals living today.

They are calling with a sacred pledge to be with us, all the way.

This is the Sacred Promise.

Yes, there are potential risks involved, and we must be responsibly mindful of them. These fall under two general categories discussed in detail in appendix B:

(1) Believing you are in communication with Spirit when you are actually fooling yourself or being played by your own unconscious pathology

252

(2) Removing a veil that may be protective of humanity and subjecting you to negative spiritual influences

However, if we do not pick up our metaphorical spirit phones
and answer the call, we will not get the message. The evidence points
in the direction of them being here and promising to help, *and* that
we must do our part as well.

Try to imagine that you are Susy Smith or Albert Einstein or
even Angel Sophia. Imagine that you are in Spirit standing behind
me right now, watching me as I am typing these words. Imagine that
you know that all of these claims, and more, are real. And imagine
that you, from the other side, are saying to me:

"Hurry up. Get the message out. We're here. We want to help
you, humanity, and the planet.

"Do what is necessary. Design the experiments; conduct the
research. We'll be there and do our part.

"And when you finish your experiments—from exploratory self-
science to formal laboratory research—whatever the results, we will
continue to assist and protect you, forever meeting our sacred
promise to help you and your precious gem of a planet to survive,
thrive, and evolve."

Meanwhile, I don't hear you saying these things to me.

Or if I do, I think, "Ah, it's just my creative unconscious, wishful
thinking, or a mind that's lost its rudder."

How would you feel if you were on the other side and got that
response after all of your efforts now and down through time?

Are you feeling frustrated? How about angry? Are you losing
patience with me? Do you feel like giving up on me and all of humanity?

Remember that I am personally an orthodox agnostic through
and through. No matter how many times I am nudged (if not
shoved) off the fence of doubt by the ever-increasing and most amazing proof-of-concept data, I soon enough pick myself up and climb
back on the fence.

AFTERWORD

It Is Potentially Dangerous to Remove the Veil and Open Pandora's Box

It may be wise, if not critical, to remember the lesson of the legendary story of Pandora's box.

If Spirit really exists but there is a veil separating it from you and me, the veil may exist for protective reasons, and it should not be removed mindlessly. Though this book does not delve into the potential risks involved in removing the veil—I do discuss them at greater length in appendix B—it is important that we be mindful of them.

As I share in *The Truth about* Medium, if our energy and information, including our consciousness, continue to live on like the light from distance stars, then this not only applies to saints but it does so to sinners as well.

If Princess Diana's and Mother Teresa's energies and information continue, then Adolph Hitler's and Saddam Hussein's energies and information continue as well. If the living energies of the son of God (believed by many to be Jesus of Nazareth) continue, then the living energies of the scourge of God (Attila the Hun) continue as well. From a purely energetic physical point of view, photons are photons, regardless of the histories of their respective energies and frequencies.

After Einstein discovered his famous formula—energy equals mass times the speed of light squared—he came to realize that it was metaphorically like a magician's wand for scientists. In the hands of responsible scientists, it could open new vistas for humanity; wielded by those in the service of the military, it could create the ultimate weapon of mass destruction. Simply stated, the magician's wand is neither good nor bad on its own; its effects depend on its user. It could be used to cure disease or to produce genetic abominations.

Just ask Evan how the Harry Potter novels show the dangers of placing a magical wand in the hands of children who lack the knowledge, skills, and wisdom to use it safely and for the highest good of all.

Though the children are not depicted as being nearly as abusive as some of the evil adults in the stories, they can seriously hurt each other as well as themselves.

The same can be said for removing the veil. Just as the science of Spirit is in its relative infancy, humanity as a species often behaves in a juvenile manner, which includes sometimes acting in dangerous ways. It is unclear whether we are as yet old enough, knowledgeable enough, and wise enough to allow the removal of the hypothesized barrier that protects us from Spirit.

However, if I am to believe what numerous research mediums are telling me, the higher spiritual reality is prepared to help protect us, not only from ourselves but also from supposedly negative energies that exist in the larger spiritual reality.

For me, a critical component of *The Sacred Promise* hypothesis is that spirits are here to protect and guide us, both here and there.

We Are Ready and the Time Is Now

So should we do the necessary research, and do it now?

The scientist in me says, "The available proof-of-concept evidence suggests that there is a real phenomenon here, but we can't know for sure unless we conduct the definitive experiments." I recall the numerous self-science (Type I) as well as exploratory laboratory investigations (Type II) and formal experiments (Type III) revealed in this book that point to this conclusion.

The clinician in me says, "If knowledgeable and responsible spirits are here and can help all of us, individually and collectively, then we need to develop ways to communicate and cooperate with them." This recalls the first of the three sacred promises introduced in the preface.

The philosopher in me says, "If the existence of Spirit is true, then I wish to live my life truthfully, and know that it is true." I remember

AFTERWORD

Yale's motto, *Lux et Veritas*, Light and Truth, and I chose to live my life this way.

The caring person in me says, "If Spirit is real, and Spirit cares about us, then I certainly care about letting it help us." I recall Susy and Shirley, Diana and Albert, Sophia and Sam, to highlight a few, and if they exist, then their wishes and dreams matter to me.

The naturalist in me says, "If Spirit can help us nurture and protect animals, plants, and the earth as a whole, I absolutely want Spirit's help." I remember the vanishing tigers, polar bears, and rainforests, and my heart goes out to them.

And the parental part in me says, "Our children and grandchildren should not have to suffer because their parents and grandparents have yet to awaken. Their hearts and souls should not be destroyed, their possibilities cut short, because we are too afraid or too selfish to act." I remember the young Evan who wished to comfort his mother by advising her to call on her guardian angels, and I suppress the tears in my eyes.

There is an old Native American saying that puts this in perspective:

"The work should honor our parents' parents' parents and serve our children's children's children."

In final analysis, this book, and its urgent call to action, is not merely for us and our souls, for the larger spiritual reality and their souls; it is for our children and their children's children and their collective souls. It is about love of self and others in the past, present, and future.

It is said that we are at a precipice, a tipping point, a time when courage must be combined with wise action to take us beyond where we have been.

There is an ancient Chinese proverb from the sixth century BC, which says:

"If we do not change direction, we will end up where we are going."

256

AFTERWORD

Einstein said something similar:

"No problem can be solved from the same level of consciousness that created it."

This is ultimately a consciousness-raising book. I am being blunt. Evan and his parents have reminded me—and hopefully you, too—that if ever there was a time for us to be courageous, to see our higher selves in the mirror, and to experience the power of greater possibility for humanity and the planet as a whole, this is it.

The journey awaits us.

APPENDIX A

Frequently Asked Questions and Commentary

The important thing is not to stop questioning.

—Albert Einstein

What led you to explore the Spirit Intention hypothesis? Was it scientific theory, laboratory experiments, or direct personal experiences?

I was led to focus on the Spirit Intention hypothesis—and its practical applications in terms of our possible relationship with Spirit—through the combination of all three, but especially the third.

As we discussed in chapter 3, I logically came to the conclusion (as many philosophers have) that it is theoretically impossible for us to prove definitively that any one of us—or any thing, for that matter, be it an animal, a plant, a machine, or a planet—has conscious experience.

Moreover, I came to accept the conclusion that it is impossible for us to establish definitively beyond any doubt that our consciousness has conscious intention or will.

The fact is that save for our own personal consciousness and intentions, which we all experience firsthand, we must only infer the existence of consciousness and intention in others. The key issue is realizing the necessity for responsible inference in both science and our personal lives.

When I began focusing on the nature of the inference process, I began to wonder how we infer the existence of intention in others. In a number of exploratory investigations as well as formal laboratory experiments, unplanned events sometimes happened, strongly implying that spirits have minds of their own.

You will recall in chapter 4 how my mother seemingly barged into the last reading in the Canyon Ranch experiment, even though she had been uninvited. Or you will remember in chapter 6 that a mysterious deceased woman showed up in the double-blind experiment, claiming that the deceased man being summoned was sleeping and she would not wake him, and subsequently learning that the information fit Susy Smith. These surprising events, which messed up our formal experiments, were successful and powerful hints that the information the mediums were getting could be inferred as coming from beings who are alive.

However, it was a collection of unexpected events in my personal life, which I observed with a scientist's mind, that were the most powerful and meaningful.

For example, you will recall in chapter 5 that Susy apparently brought a mysterious deceased woman to a medium, which served as a spiritual healing experience for her sister and parents. This observation, which was replicated many times, led to my formulating the double-deceased research paradigm.

It is my fervent hope that systematic research addressing the Spirit Intention hypothesis will be conducted—sooner rather than later—in multiple university laboratories.

If the promise of this book is fulfilled, such research will not only demonstrate that beings like Shirley and Susy have at least as much consciousness and intention as you and I do but hypothesized angelic beings such as Angel Sophia and Archangel Michael do so as well.

Why did you reveal so much of your personal self-science experiences in this book? How do your private life observations complement and extend your professional academic observations?

Why include so many private-life observations? The reason is simple: this is where the majority of the most innovative and important inspirational events for the work happened.

Together with the exploratory investigations and formal laboratory experiments, they provide compelling proof of concept of the premise that Spirit exists and can play an important role in our individual and collective lives.

They offer proof of possibility that all this, and more, may be real.

As a general rule, scientists typically place little stock in personal anecdotes, even when the accounts are experienced and witnessed by trained scientists applying the scientific method in their daily lives. Scientists appreciate that it is virtually impossible to generalize from single-case studies or even collections of case studies. Case studies, no matter how carefully observed and analyzed, and even if they follow the scientific method, fall short in terms of drawing any firm general conclusions.

However, personal anecdotal experiences like those carefully observed and reported in this book can serve three important functions:

1. They can stimulate the raising of innovative and important questions, which can advance our envisioning of expanded possibilities and connections in nature and the universe.

2. They can uncover new, if not seminal, observations that inspire future controlled laboratory research.

3. They can demonstrate how the observed phenomena apply directly to real life and therefore can be useful for us individually and collectively.

As you probably guess, I have done my academic homework in this area. I have read much of the history of mediumship research, as well as various collections of case studies documenting striking and meaningful afterlife communications. I recently wrote the foreword to Josie Varga's book *Visits from Heaven*, which is a collection of afterlife accounts from around the globe and backed by the testimony of scientists and clinicians investigating such communication.

However, when personal accounts (private-life observations) are collected and analyzed in a scientific manner, and they are integrated with proof-of-concept exploratory investigations as well as formal laboratory experiments (professional-life observations) that complement and support them—as modeled in *The Sacred Promise*—the combination becomes compelling in terms of pointing to a genuine proof of serious possibility, if not serious probability.

As Stephen Sondheim wrote in his musical *Sunday in the Park with George*, about the life of the famous pointillist painter Georges Seurat: "A vision's just a vision if it's only in your head; if no one gets to see it, it's as good as dead. It has to come to life."

By sharing a scientist's visions—in my case, numerous thought experiments and associated questions—combined with personal experiences and preliminary laboratory investigations and formal experiments, I hope this important information has come to life for you.

And if you happen to scoff at the genuine possibility of the existence of Spirit and you dismiss this book out-of-hand, I hope you realize that you may be ridiculing conscious beings (in Spirit) with real feelings just like you and that while your opinions should be respected, your derision is neither respectful nor kind to them.

Are there risks associated with attempting to connect with Spirit?

The answer is yes, though many people prefer not to think about them.

I briefly mentioned risks associated with removing the veil in the afterword. I have included a more extensive discussion of risks versus rewards in appendix B. This includes the risk of believing that you are communicating with your guides, when in fact you are actually fooling yourself.

Discerning the difference between genuine communications versus imagined interactions is one of the greatest challenges in this work.

What has convinced you that the observations suggesting the existence of spirits and their ability to play a guiding role in our lives cannot be explained simply as illusions of the mind, and how does self-science protect us from drawing illusory conclusions about such observations?

In order to address this important question, it is essential that we understand the process of discerning the difference between what is a real correlation and what is an illusory one. This discussion gets a bit technical; you might want to read it slowly. Moreover, I have written appendix C, which discusses this issue in greater depth.

APPENDIX A

Briefly, the term *illusory correlation* was coined by Professor Loren Chapman and colleagues at the University of Wisconsin and actively researched by them in the 1960s. What Dr. Chapman and his colleagues discovered was that clinical psychiatrists and psychologists were prone to make certain errors of inference—and therefore came to false conclusions—when they simply observed information exchanges in their clinical practices and did not record them systematically.

When undergraduate students were presented with the same kinds of clinical information, and they were not forced to accurately record their observations, they reached the same false conclusions. In other words, when people rely exclusively on their memories, they end up drawing certain conclusions that are subsequently discovered to be illusory, versus those when recorded and examined via systematic research.

Moreover, Dr. Chapman and his colleagues found that professionals and students alike, when left to their own devices, consistently missed real correlations and failed to discover true patterns in the data when the relationships did not fit their expectations or beliefs. They literally missed the truth.

As mentioned above, a more complete description of illusory correlations and how to avoid them, including my actually witnessing the process of forming illusory correlations when I took a graduate course on psychological testing with Dr. Chapman in the spring of 1967, is provided in appendix C.

What self-science, or at least my methodology, does is require us to keep accurate records and actively explore alternative potential correlations or relationships in our own lives.

It encourages us to be open to possibilities (including surprises) and to not prematurely jump to conclusions that might be illusory.

Because I have conducted both self-science (personal life observations) and laboratory science (professional life observations) and integrated the two, I feel that I can see the bigger picture. From this perspective I can draw the firm conclusion that the illusion hypothesis does not, and cannot, explain the totality of the observations reported in this book.

How do you respond to professional skeptics who debunk this kind of research and will probably label you as unscientific or weird, if not crazy?

In all my books I have attempted to take a skeptical perspective—meaning a wondering, questioning, and critiquing point of view. Being a questioner, which I have referred to as being an orthodox agnostic, is in my blood, if not my soul.

However, there is a difference between skepticism and cynicism, between genuine wondering about psychic occurrences and dogmatic disbelief about them, no matter what the evidence—which is another way of describing an ardent belief that something is not real or is impossible.

I discuss the fundamental difference between healthy and unhealthy skepticism in appendix D. Some of the language used in that appendix is frankly negative and feisty—this is unfortunately unavoidable, because I must use the extreme language of certain professional superskeptics to make the take-home points clear.

Meanwhile, I remember the wise words of Rama, a character in John Nelson's visionary novel *Matrix of the Gods*. He said to a skeptical female reporter (who worked for a cynical man):

> I am sure you will do your best to be fair, and I wish you the best of luck with it. I can imagine how strange all of this must sound to you, to your audience, and to those you answer to. I consented to the interview knowing full well what use could be made of it. But, ultimately, your designs, or those of others,

including myself, are mere shadow plays. I have done what I could; do what you can do, and let it go.

Writing this book was one way I could do what I can do.

Can spirits, if they are really here, help us heal the world and ourselves?

Of course, this is the sixty-four-trillion-dollar (if not gazillion-dollar) question. The premise of this book is that the answer to this question will be yes. However, from a scientific point of view, this premise is a hypothesis in evolution, supported by exploratory observations and preliminary investigations and experiments.

I believe that in light of humankind's and the planet's current plights, it is essential that we address this question as quickly, creatively, and responsibly as we can. And we need all the help we can get.

As I brace myself for the potential onslaught of emotion and criticism concerning *The Sacred Promise* and its implications, I remind myself of Michael Jackson's inspiring words in his song "We Are the World," where he writes about "hearing a certain call" and "lending a hand to life."

If you, the reader, "hear a certain call" and can "lend a hand to life," I thank you, and I smile.

How will the needed future systematic research be funded?

To address this challenging question, a bit of history is helpful.

In 2005 a well-known television producer approached me about the possibility of participating in a series on the training and development of mediums. Her idea was to select a group of highly motivated novices—people who wanted to become mediums—and have them work with a group of experienced research mediums. The goal was to chart their ability to learn more accurately and meaningfully how to obtain information from Spirit with both personal and scientific perspectives. My role was to provide the scientific testing of the mediums in training.

The prospective television series had been developed to the point of a contract being prepared to produce the pilot show. However, the president of the network at the time, who had enthusiastically supported the series, was unexpectedly relieved of his duties and a new team of executives was hired.

I was disappointed (and a bit disheartened) by the failure of the series to materialize. I had hoped that it would provide the means for collecting comprehensive data on the process of novices developing medium skills and using this information for personal as well as professional purposes. To the best of my knowledge, no formal research has been conducted on the training and development of connecting with Spirit or on the broader question of the reality and role of Spirit in our individual and collective lives.

Even in the best of economic times, conventional funding sources—such as the National Science Foundation or the National Institutes of Health (both of which have funded my mainstream scientific research in the past)—are not open to supporting such challenging and controversial research. At the present time the promise of this research is being funded almost exclusively by the private sector.

The media might be one viable means of fostering this work because of the potential win-win-win-win-win-win (yes, six wins) nature of such collaboration:

1. It would be a win for the network, because they would make money in airing such a series, as well as do something positive for humanity.

2. It would be a win for the producers, because they would make money in doing the series, as well as do something positive for humanity.

3. It would be a win for the mediums, because they would earn money participating in the series, as well as gain exposure for their work and do something positive for humanity.

4. It would be a win for the scientists, because we would be able to conduct research on the question and hopefully publish the findings, as well as do something positive for humanity.

5. It would be a win for the public, because we would be able to view something that was entertaining, educational, meaningful, and inspiring, and which may inspire more of us to do something positive.

6. It would be a win for alleged spirits, since their voices would be more widely heard and potentially taken seriously, so that they can help us better our world.

The truth is, until political as well as scientific leaders wake up to the promise of this work and reasonable public funding becomes available, advances will come slowly and piecemeal, and progress will be neither systematic nor programmatic.

Someday, and hopefully sooner rather than later, the public and private sectors will join forces and do for a future spirit research program what we previously did for the space program. Together we will make a serious collective commitment of resources and talent to take us to inner space, much as we did to explore outer space.

Meanwhile, a small group of people—some trained in science, some not—are actively working behind the scenes in a few universities as well as private institutions to explore these possibilities.

The truth is that I get discouraged from time to time about seeing this work move forward. When this happens, however, I am reminded of Clarissa and her seemingly persistent guides, as well as Evan and his parents' need to know if guides are real. I then re-remember why we and others are doing this work, and I smile again.

What does the Great Spirit or God have to do with all this? Can S/He provide the ultimate guidance we need in our lives?

These are profound scientific as well as spiritual questions. My working hypothesis is that the answers to these two questions are everything, and yes.

My conclusions about the probable existence of a Universal Intelligence and Source of everything existing in the universe, including within us, are developed in detail in *The G.O.D. Experiments*. In that book I explained how science slowly but surely took me to God.

Included in *The G.O.D. Experiments* was a poem that came to me. It is the only poem that "I" ever wrote. I say "I" here because the poem you are about to read appeared in my mind, fully formed, on an airplane as I was traveling to the East Coast. I was literally high in the sky when I experienced it (and for the record, I had not consumed any liquid spirits on the trip).

I wrote down as much as I could remember of the poem (which was most of it), and then tweaked it a bit for clarity and rhythm.

For example, when I first put the poem on paper (I literally wrote it out by hand), the next to last line read, "Enlightened compassion, the flight of the dove." I had no idea what this meant at the time, so I later tweaked it to read, "Awakened compassion, a Gift from above?" If the original line is more poetic and has greater spiritual meaning than my attempt at editing, we can attribute the original to something other than my logical mind.

APPENDIX A

The poem expresses the parallel relationship between information and energy on the one hand (in science) and soul and spirit on the other (in spirituality). I have included the poem below, partly because of its direct relevance to *The Sacred Promise* and partly because it reminds us that there is a deep parallel that exists between science and spirituality and is increasingly coming into focus in the twenty-first century.

Though each of the individual stanzas in the poem is easy to read, their meanings are quite deep. You may find it helpful to reread the poem and ponder each stanza with a questioning mind and a caring heart.

Is the soul like information, and is spirit like energy?

I suggest that these words, expressed in this complementary fashion, provide us with a simple yet integrative theoretical framework for rejoining science and spirituality.

Soul as Information, Spirit as Energy
Received and edited by Gary E. Schwartz

What, pray tell, are Spirit and Soul?
Are they one and the same?
Are Soul and Spirit a functional whole?
Derived from a common name?

Or is it the case that Soul and Spirit
Reflect two sides of a coin?
Where Soul reflects the Information that fits,
And Spirit, the Energy that joins?

Is Soul the story, the plan of life?
The music we play, our score?
Is Spirit the passion, the fire of life?
Our motive to learn, to soar?

Soul directs the paths we take,
The guidance that structures our flow.
Spirit feels very alive, awake,
The force that moves us to grow.

If Soul is plan and Spirit is flame,
Then matter is alive, you see.
Nature may play a majestic game,
Of information and energy.

I'd love to believe that wisdom and joy
Reflect God's plans and dreams,
That Soul and Spirit are more than toys,
And both are more than they seem.

Could it be that the Soul of God
Is the wisest of plans, so grand?
And the Spirit of God is the lightning rod
That inspires God's loving hand?

Could Soul be wisdom, and Spirit be love?
Together, a divine partnership?
Purpose and passion, a duet from above,
The ultimate relationship?

The relationship of Spirit to Soul,
So simple, profound this team.
For Spirit and Soul the ultimate goal,
To understand this theme.

Soul as wisdom, Spirit as love—
Information and energy;

Awakened compassion, a gift from above?
Someday, pray tell, we'll see.

You may wonder, as do I, was the essence of this poem given to me by the energy of the Great Spirit?

Could this have been a special moment, high in the sky, when I briefly received information from "The wisest of plans, so grand" and experienced "the lightning rod that inspires God's loving hand"?

Has the author "lost his mind," or has he "found his soul"?

As Einstein reminds us, the important thing is not to stop questioning . . .

APPENDIX B

Risks and Rewards of Seeking Spirit

Take calculated risks. That is quite different from being rash.
—General George S. Patton

Though the inherent promise of seeking Spirit is generally positive—be it in the laboratory or in our daily lives—the search does have its risks for everyone concerned. The risks range from believing you are in communication with Spirit when you are actually fooling yourself, to removing a veil whose presence may reflect higher wisdom and be protective of humanity and the planet.

Besides teaching me to be prepared for surprises, this research has reinforced for me the important lesson of balancing the process of exploration and discovery with caution and wisdom. This includes remembering the lesson of Pandora's box (mentioned briefly in the afterword) and being mindful of the need for critical discernment at every level.

Unlike a scalpel that is basically neutral—whether it promotes curing or killing is up to the person using it—spirits are not necessarily neutral, nor are the people seeking contact with them.

How does a given person know, for sure, if she is receiving communication from Spirit (or spirits) rather than listening to their creative unconscious or some other source of the information?

How does a given person determine if he is accurately detecting and interpreting the information, regardless of its source?

And how does a given person ascertain whether the source of the information, if it is spiritual, has healthful or harmful intentions?

Let me offer a few examples, both personal and experimental, that illustrate these risks and challenges.

Please understand that I mean no offense or criticism here to any individual. My purpose is to raise questions. Whether you are just beginning the process of connecting with Spirit or you are a professional intuitive whose livelihood is related to your connecting with spirits and higher guides, it seems prudent to consider some of the potential risks as well as celebrate their probable rewards.

Just as Einstein's formula linking energy and matter was associated with serious dangers as well as great advances, connecting with Spirit has its possible negatives as well as obvious positives.

Can We Make Up Spirits in Our Heads?

Can we be fooled into thinking we are receiving communication from Spirit when we are not? The answer is unfortunately yes.

For example, I have a fairly vivid imagination—not in a visual sense but rather in a verbal or abstract sense. I can hold imaginary conversations in my head, and at times they can seem quite real.

Of course, when I am playing this imaginary game, I am acutely mindful that this is my intention, and I assume correctly (or incorrectly) that I am in fact having the conversation with myself.

I can also ask questions of the Universe in my head, as I shared in my book *The G.O.D. Experiments*. I have discovered that new information frequently appears in my consciousness when I do this. Sometimes it is highly accurate (and surprisingly so).

As an illustration, I once asked the Universe for another name for God, and I immediately heard the name Sam. I subsequently learned that Sam, short for Samuel, comes from the Hebrew *Shemuel* which happens to mean "the Name of God."

Not only was I ignorant of this semantic fact at the time, but I discovered that most people—save for those who know the Hebrew language—are.

When I ask questions of the Universe, I cannot tell whether the information is coming from:

1. My imagination

2. My unconscious

3. Someone else's mind (a form of unconscious telepathy)

4. Jung's hypothesized "collective unconscious"

5. Spirit

6. Divine Source

All I know for sure is that the information appears in my mind and that I can later determine whether it is accurate or not.

Hence, I am assiduously careful to be humble about attributing the information to any specific source.

You may recall that Clarissa, the intuitive in training featured in chapter 15, says that asking for information is like catching a fly ball in her mind at night and not necessarily knowing where the ball has come from, nor who (or what) has hit (or propelled) it. All that Clarissa knows for sure is that when she puts her "mental hand" up (my term), she often catches information, and sometimes this information is surprisingly accurate and useful.

In preparation for writing this book, I tried a new mind game, and the results were startling.

I discovered that if I imagined creating a fictitious deceased person in my mind, this mental process could seemingly take on a life of its own.

I could effortlessly engage in a playful dialogue with the imaginary (invented) being/Spirit, and it gave me all sorts of information whose unique constellation did not match anyone I knew or could locate on the web.

I later learned from my editor that Dr. Carl Jung had discovered a similar process, which he called active imagination.

Novelists often report that their imaginary characters can take on a life of their own as the authors craft their books. They will sometimes talk about the information flowing and at times even feel as if they were taking dictation from their characters.

I could also imagine creating a fictitious council of spirits, and they too could take on a life of their own.

For example, I once imagined a council and then asked the group its name.

I first heard, "The Council of Illusion," and then I heard "The Council of Wisdom."

I then mentally asked this imaginary council where the members came from, and I heard in my mind the galaxy Nepia. I presumed what I heard must be nonsense.

Out of curiosity, I went to Google and typed in "Nepia + galaxy." One listing was as follows: "Nepia Pohuhu, a Wairarapa adept from the Maori tribe of New Zealand, [...] stated that Matariki (the Pleiades) was a young brother of Tongatonga, and that Matariki was conveyed to the Paeroa o Whanui (another name for the Milky Way) to take care of the *whanau punga* (stars), lest they be jostled by their elders and so caused to fall."

I wondered whether this was part of Maori mythology, and if the peculiar name for the Milky Way was from their language. Could it be, as strange as this might seem, that my imaginary council was somehow associated with their stellar mythology, since many indigenous tribes claim a connection to the Pleiades?

What was remarkable to me was how real all this seemed in my head at the time it was occurring.

What I mean by real here is that the imaginary communication process seemed as fluid as my own conversations with myself in my head.

They—those of the Council—seemed as real as the imaginary me in my mind.

Another way of saying this is that I could not readily discern an experiential difference concerning the process of me conversing with me from me conversing with them.

Did I experience a difference in the content? Of course.

Did I experience a difference in the feeling of the process of the communication? No.

Whereas healthy (and sane) people can explore these similarities and differences in their minds, just as I did, people suffering from schizophrenia cannot make this distinction. Schizophrenics simply assume that the voices they are hearing in their head must be real, as in separate from them. People who are insane do not carefully question or analyze such experiences as we are doing here.

I have witnessed a number of individuals who have learned to channel who go through three stages in the development of their conscious awareness of communication with alleged Spirits:

1. Questioning whether what they are hearing is coming from their minds or from an independent entity

2. Developing the habit of consistently interpreting the experiences as coming from their guides

3. Identifying with their guides so that they cannot tell where they end and their guides begin

Unfortunately, once you go down the path of losing the ability to distinguish the difference between your imagination and something (or someone) else's, you run the risk of your own imagination getting the better of you, and you can end up sometimes erroneously labeling your mind's information as coming from an outside source.

This is a slippery slope and a potentially dangerous mental habit to develop.

Similar to developing a smoking habit, what we might metaphorically call a "spiriting" habit appears to develop and becomes very hard to break.

Some intuitives and psychics may become ungrounded and flaky as they lose their ability to distinguish between their own thoughts and those which purportedly come from elsewhere. Moreover, they may become defensive when they are questioned or challenged concerning their interpretation of the source of a given piece of information.

I doubt a university human subjects committee would allow a scientist to perform controlled research testing where people could be taught to develop imaginary spirits and then chart what happens to their mental relationship with their imaginary spirits over time. The possible risks to the subjects outweigh the potential gains to science and society.

I mention this hypothetical experiment as a thought experiment. This is a theoretical experiment; it is not a practical or advisable (read: ethical) one.

However, if someone is going to try to develop personal communications with spirits, he or she should be cognizant of the fact that there are inherent psychological risks involved in doing so. It would be unwise and irresponsible to speak of the Sacred Promise without also considering the potential negatives involved.

One way to minimize this particular risk is to remain diligently mindful of the fact that in the absence of definitive evidence about the source of whatever information you experience, you should consciously employ phrases such as "I am experiencing X images" or "I am hearing Y thoughts," rather than automatically saying, "My guides are showing me X" or "My guides are telling me Y." The former phrasing keeps you focused on the information rather than on the presumed—and potentially erroneous—interpretation of its presumed origin.

As quoted in the introduction to this appendix, General Patton reminds us that taking calculated risks is one thing, being rash with the information is quite another.

The Lesson of a Potential Failed Experiment

There are significant risks to doing certain kinds of proof-oriented research on the individuals potentially communicating with spirits.

The primary risk for the participants is that if the results turn out to be negative—i.e., it fails to support a person's expectations and predictions—it has the potential to be the kind of evidence that could be interpreted as seriously challenging an individual's core beliefs.

Having one's core beliefs challenged is difficult enough; to have one's most intimate inner experiences questioned can be even more personally threatening.

Let's try the following thought experiment.

If you are willing, allow yourself to imagine that you are a professional intuitive.

Imagine that you make your livelihood from connecting with Spirit and giving personal advice and counsel.

Imagine that you are good at this—meaning you often get accurate information about people and their deceased loved ones, and you provide practical information that helps others in their personal and professional lives.

Now imagine that you decide to participate in an experiment testing one of your core beliefs—that you can accurately communicate with your Spirit guides, and they with you.

In your case, let's imagine that the Spirit guide is your deceased grandmother.

Then imagine that you team up with another successful intuitive who also believes that he can accurately communicate with his Spirit guides.

In his case, let's imagine that the Spirit guide is his deceased grandfather.

The two of you are convinced not only that you can accurately communicate with your respective grandparents, but that they can communicate with each other on the other side.

In active collaboration with your respective alleged grandparent guides, you design (as a foursome) an ideal experiment:

1. You determine you are looking at a meaningful picture.

2. You ask your grandmother to pass this information to your research partner's grandfather.

3. Your partner then contacts his grandfather and asks his grandfather to share what your grandmother told him.

4. Your partner will accurately describe the picture you saw.

In other words, by using spirit-to-spirit communication, your partner would be able to accurately read your mind, albeit indirectly. We can call this the spirit-to-spirit assisted telepathy experiment.

Now, let's imagine that you and your partner are given the freedom to conduct the experiment at times optimal for each of you, and you repeat the experiment multiple times—sometimes you are the sender and sometimes you are receiver; i.e., your partner is looking at a picture and asking his grandfather to pass the information to your grandmother.

Finally, imagine that you are convinced that this experiment should work, and to your surprise, it does not. In fact, it fails miserably.

Think about this.

How would you feel if this experiment—the very experiment you supposedly designed with your deceased grandmother—failed to reveal reliable evidence of your purported ability to receive accurate information from her?

What conclusions would you draw?

How might it impact your business and your reputation if it was known that you tried this experiment and that it failed?

Remember, we are imagining that you are a successful intuitive; that you often get accurate information that is meaningful and useful for your clients.

Let's be sure we are clear here. The experiment you and your alleged team designed was *not* testing your applied ability as an intuitive to get this useful information for your clients—it was testing your theory about how it all works; i.e., you are getting this information specifically from your guides.

There are various possible reasons why your hypothetical spirit-to-spirit assisted telepathy experiment might have failed.

The two of you might have been stressed in the process of carrying it out and had difficulty connecting with your guides.

Another possibility is that maybe your guides wanted the experiment to fail so that you would learn a different lesson.

The fact still remains that this was your team's experiment. You and your purported grandparents designed an experiment that you firmly believed should work.

Would this imaginary thought experiment, if it actually failed, raise a question about your explanation for why you are successful as intuitives, such as your conviction that the accurate information you get for your clients comes specifically, if not exclusively, from your guides?

The answer logically is yes.

This is actually a relatively mild example. There are other more bold and proof-oriented experiments that can be designed, which if they failed would logically raise even more doubts about your explanation regarding the source of this information.

The issue here is not whether intuitives who believe in the existence of guides are ultimately right or not—future experiments might (or might not) validate their explanations. And the issue is not the risk to your belief that the spirit-assistance conclusion is correct in the general sense, but that it applies to you in a specific sense.

Common sense dictates that to minimize the risk of participating in a proof-oriented experiment that might challenge or even disprove your core beliefs—by failing to confirm your prediction—the safest thing is to not volunteer to do such research in the first place.

Of course, fear of discovering that you might be wrong is ultimately not a wise way to live one's life. We will return to what Dr. Carl Sagan calls "the heart of science" at the end of these appendices.

Is It Safe to Remove the Veil?

I have been privileged to know and work closely with numerous professional sensitives, intuitives, psychics, mediums, channels, and healers. Virtually all claim that, with proper training and good intentions, it is safe to receive information from the larger spiritual realm.

The key phrases here are *proper training* and *intentions*. Responsible professionals who believe they are in contact with Spirit are, as a general rule, mindful that there are negative energies or spirits as well as positive ones, and that it is essential for them to keep the potentially harmful ones at bay.

The sordid history of children and adults playing with Ouija boards is replete with stories—real, exaggerated, or imagined— of dangers. The risks are typically associated with giving malevolent spirits access to the minds of the people using the boards.

To a person who doesn't believe in the existence of a greater spiritual reality, this issue is moot. Moreover, the skeptic would interpret the behavior of a serial killer, for example one who claimed that he or she was following the guidance from a specific evil spirit, the devil, or even God, as engaging in a pathological self-deception.

However, if the existence of Spirit is real, and if the energies and consciousness of benevolent beings exist by the same mechanisms as the energies and consciousness of malevolent ones, then there is a serious risk in us opening ourselves to Spirit, especially in the absence of appropriate training and intentions.

Professionals who believe they are in regular contact with spirits are typically taught various protective techniques, from taking Epsom salt baths to practicing specific meditations and prayers. (To the best of my knowledge, there has been no formal research investigating what effects such practices have on the physical, psychological, and spiritual health and well-being of people connecting with Spirit.)

We live in times where ghost hunting in the media is at an all-time high. Curiously, while writing the first draft of this appendix, I received a surprise email from Christopher Robinson. He is the precognitive dream detective from England whom I wrote about in *The G.O.D. Experiments*. He informed me of his then upcoming participation in a popular ghost-hunting TV show called *Most Haunted*.

Was this a synchronicity? According to the website promoting the eight-episode series, one episode included a live séance bringing "a dark criminal undertone to the event." The séance attempted to bring forth killers involved in "gruesome murders, hordes of brutal hangings, and twisted torture."

I mention this particular show for one reason: it was focusing specifically on evil and was implicitly encouraging the public to go hunt for it!

Is it wise for humanity to seek out evil?

There is an essential protective reason for placing dangerous individuals behind bars—to keep them from harming others. This is common sense.

And it is equally common sense to realize that if malevolent spirits do exist, then there is a similar protective reason for preventing them from harming us or using us to harm others.

Though most believers don't like to think about this downside, it is prudent, if not essential, that we are open to the possibility of an extreme risk in naively reducing or removing the barrier between the physical and the spiritual.

As mentioned previously, just as it is not wise to give young children magic wands (or knives)—since without proper training and intentions, they can be dangerous for children and those around them—it is probably not wise to give people general access to Spirit without proper training and intentions.

In many ways we are like young children, at least when it comes to connecting with Spirit.

Various scientists, including me, see important theoretical and practical consequences of being able to develop a communication technology between humanity and Spirit. If I did not envision these positive benefits, I would not attempt to pursue this work, nor would I have written this book.

At the same time, however, I am mindful that if Spirit exists, there are serious potential risks in giving it increasing direct access to us. The risks not only come from spirits; the danger involved comes from us as well.

In *The Afterlife Experiment*, I discussed the theoretical possibility that if a soul phone could be invented, individuals here in the physical might employ unscrupulous spirits to spy on people and use this information for malevolent military purposes, or worse. I envisioned the possibility of the CIA being expanded to become what might be termed a DIA—a Deceased Intelligence Agency. Of course there are benevolent (preventive) as well as malevolent potential uses of spirit-assisted spying.

We could elaborate on other potentially threatening, perhaps terrifying, risks associated with the naive removal of the apparently protective veil.

My purpose here is not to encourage panic, nor to imply that we should cease and desist from the path of making increasing contact with a greater spiritual reality. My intent is to encourage all of us to see the big picture—the risks as well as the rewards—and to pursue this sacred work with caution in addition to creativity.

I can say with great assurance and experience that the majority of individuals I have worked with—ranging from medical intuitives and mediums to channelers and spiritual healers—are among the most loving, thoughtful, and responsible people I have known. Though I question many of their specific beliefs about Spirit, as well as some of their professional practices, on the whole they seem to be kind and caring people.

On the other hand, I have known a minority who evidence negative qualities of naiveté, egotism, self-centeredness, self-aggrandizement, irresponsibility, self-delusion, and even overt deception (and I have on rare occasions paid a professional and personal price for this). And I have known individuals who use alcohol or drugs to reduce their stress and cope with being hypersensitive.

The truth is that it is not easy being a professional spirit communicator, and I try to celebrate all individuals who seek spiritual connection for personal and collective growth and evolution.

But we are metaphorically like children when it comes to connecting with higher spiritual realms, and as a species we have a lot of growing up to do.

From the Heart of Science to the Heart of Wisdom

One of my scientist heroes, Dr. Carl Sagan, taught me a deep lesson about what it means to be a true scientist.

He wrote: "When Kepler found his long-cherished belief did not agree with the most precise observation, he accepted the *uncomfortable fact*. He preferred the hard truth to his dearest illusions; that is the heart of science."

The essence of a true scientist is being a truth seeker. However, few things are more difficult than to *accepting uncomfortable facts* as real and then choosing these hard truths over our dearest illusions. My experience is that scientists, as a rule, are not necessarily more open to doing this than anyone else.

Depending upon what your cherished beliefs happen to be (i.e., whether you are a believer or a disbeliever in Spirit), you might interpret the following two statements as being uncomfortable facts.

As a thought experiment, let's imagine, just for the moment, that both of these statements may be true:

1. Some, if not all, of what you experience as communicating with spirits could be your own unconscious.

2. Negative as well as positive spirits exist, and there are significant risks in connecting with Spirit.

The question arises, how and when do we decide to change our minds as a function of discovering new facts, especially if the facts are uncomfortable and challenge our most cherished beliefs?

What I have learned in the process of doing this research might be viewed as an extension of Dr. Sagan's description of the heart of science.

I suggest that the heart of wisdom might be defined as "knowing how and when to change our minds as we come closer to truth, and then doing so." In other words, the heart of wisdom is not merely knowing how and when to change our minds but also having the courage and strength to actually change them.

If any area of human exploration requires an integration of the heart of science with the heart of wisdom, it is the area of spirit-assisted living.

Balancing and Integrating Sensing, Thinking, Feeling, and Intuiting

To foster the heart of science with the heart of wisdom, it is valuable to ponder Dr. Carl Jung's concept of transcendental functioning—what might be termed balanced integrative knowing. There are two fundamental ways that we gain information:

Sensing—what happens in the world that we can experience with our physical senses

Intuiting—having impressions, hunches, gut feelings, spontaneous images and thoughts that seem to come from "elsewhere," e.g., from Spirit

And there are two fundamental ways we process that information:

Thinking—using logic and reasoning to select, process, and interpret information

Feeling—experiencing our emotions in a given situation, be they positive or negative

Dr. Jung proposed that all four modes of gaining and processing information are meaningful and important and that no one of them should be viewed as primary or necessarily always correct.

Scientists as a rule tend to place priority on sensing and thinking; intuitives focus more on intuition and feeling.

The greatest challenge and opportunity for human evolution is for us to combine all four modes of acquiring and processing information. We can learn not only how to integrate them but how to discern which ones to follow and when.

From Training Wheels to the Tour de France— Honoring the Sacred Promise

From both a scientific and a wisdom point of view, we could say that as a species we are receiving our first children's bicycle with training wheels.

Let's consider this spirit-seeking-as-bicycle-riding metaphor more thoroughly.

There are obvious risks associated with riding a bicycle. Most of us have fallen down on one or more occasions, receiving various degrees of injuries in the process. However, most of us eventually graduated from four-wheelers and learned how to ride two-wheelers. Many of us have discovered that bicycle riding can be fun.

Some of us have advanced to mountain cycling, which takes substantially more skill as it exposes us to significantly more risks.

Though I have not tried the sport (given my age and physical condition, it would be rash for me to do so), I have been told that mountain biking is thrilling and can be quite an adventure.

And a select few of us have reached the pinnacle of cycling and even competed in the Tour de France.

Is riding a bicycle—be it on a child's four-wheeler or on a speed racer's two-wheeler—worth the risks? Most people who ride bicycles say yes.

The same applies to seeking Spirit. People who attempt to connect with Spirit seem to believe, more often than not, that it is worth the risks.

However, we can ask the deeper question: is it possible that we were, so to speak, as a species "born to ride"?

Is seeking Spirit in our genes?

My working hypothesis is yes. The journey to Spirit awaits us. This is the Sacred Promise.

To quote *Contact*, one of my favorite films, "Do you want to take a ride?"

APPENDIX C

Are Spirits Illusory and Are We Fooling Ourselves?

Skeptical scrutiny is the means, in both science and religion, by which deep insights can be winnowed from deep nonsense.

—Dr. Carl Sagan

The human mind is truly gifted at fooling itself.

I am especially mindful of this cognitive capability and I do my best to minimize it. Are we being fools? And even worse, are we fooling ourselves if we come to the conclusion that spirits are real and can play an important role in our individual and collective lives?

My strongest detractors rarely appreciate the fact that I have a deep understanding of the potential for self-deception, and I am particularly vigilant about its possible influence. A major reason why I am ever mindful is because my students and I have conducted two decades of systematic laboratory research—published in mainstream scientific journals—on the psychology and psychophysiology of self-deception and its effects on mental and physical health.

No one likes to be called a fool. Skeptics in particular have an aversion to being seen as foolish by themselves and others.

I learned the lesson of the self-deceptive mind early in my academic education; the example was provided by Professor Loren Chapman at the University of Wisconsin in the spring of 1967. I could never have guessed at the time I took his course on psychological testing that it would offer a key to dispelling one of the greatest fears associated with the scientific investigation of the spirit hypothesis. Though I subsequently transferred to Harvard, Professor Chapman's inspiring course left an indelible and invaluable mark in my mind and heart.

My awakening to the parameters of self-deception began with Dr. Chapman's lecture on the topic of the draw-a-person test. I invite you to relive with me this experience so you can deeply appreciate Dr. Chapman's take-home message and its direct application to *The Sacred Promise*. Once you have shared this experience with me, you will be able to see illusion and self-deception in a new light.

The Professor Deceives His Trusting Students

I was sitting in a classroom with approximately sixteen fellow PhD students in psychology and related fields, and my professor requested that I take a research test. Since I was interested in the topic and wanted to get a good grade, I was more than willing to participate.

The professor explained that he would be presenting slides of drawings made by patients suffering from different mental disorders, including depression (extreme sadness) and paranoid schizophrenia (extreme illusory fear). An example of a psychotic's paranoid fear might be that secret CIA agents were trying to kill him, when in fact this was not the case based on clear and convincing evidence to the contrary.

Patients were asked to draw their pictures. (In the 1940s–1960s, the draw-a-person [DAP] was a routinely used mental health projective test.) The professor explained that at the bottom of each drawing was the actual diagnosis of the patient as obtained from interviews and other psychological methods.

He explained that we would be seeing a representative sample of forty different pictures projected on the screen one at a time. Our task was to discover if there were any patterns that appeared between specific aspects of the drawings and the patients' diagnoses. He explained that we were not to take notes and quantify the findings during the test but instead to use our nascent information-processing abilities to discover possible relationships. Our goal was to be exploratory, to be psychological detectives.

We viewed each picture, one by one, and began to form impressions about possible relationships. After viewing the complete set of drawings, we wrote down our initial impressions. I noted that those that featured big eyes seemed to be created more frequently by people diagnosed with paranoia. When the class was surveyed, we were surprised and pleased to discover that approximately 80 percent of the class had come to this same conclusion. The professor then congratulated us and told us that we had rediscovered what skilled psychiatrists and clinical psychologists had (1) consistently observed independently in their clinical practices and (2) regularly reported at clinical meetings.

At that moment we felt relieved and even a bit self-satisfied. Then our bubbles were burst, and none of us saw the reversal of expectations coming.

The professor went on to explain that when scientists actually attempted to verify whether the big-eyed–paranoia hypothesis in the drawings was indeed correct, the research continually failed to support the hypothesis. In other words, the controlled research never replicated what the clinicians were convinced was true.

How could this be the case?

Not only did skilled clinicians come to this conclusion, but also we had just done so. We had seen it with our own eyes.

And yet, systematic research failed to support this conclusion. At that moment I was startled and confused.

Then the professor dropped an intellectual bombshell.

He confessed that he had not told us the whole truth.

At first I could not believe what I was hearing, and it shook me to my core.

Yes, the drawings were made by patients. And yes, the diagnoses were obtained from actual interviews and other methods.

However, our professor had lied to us about one critical fact. The labels at the bottom of each picture were not actually those of the patients who drew them. Instead, the labels had been randomly assigned to the pictures.

That's right: they were randomly assigned. There was actually no empirical relationship between big eyes and paranoia in these pictures. The connection we thought that we had detected between them was in fact not true.

Many of us could not believe this about the pictures.

Was the professor pulling our legs about all this association?

I wondered: Which was the real deception—the professor's claim of a random assignment or our own perception? Had the professor deceived us or had we deceived ourselves?

The professor requested that we review each of the pictures, but this time keep track of our individual hypotheses and see how often the data actually fit our initial impressions.

When the pictures were shown again, we took this data-collection task very seriously since we were clinical psychology research scientists in training. To our amazement, we discovered that the professor was correct. There was indeed no relationship between big eyes and paranoia. In other words, there were just as many drawings of big eyes with other diagnoses as with the diagnosis of paranoia.

Some of the students became quite upset as they discovered this factual assessment. A few were convinced that they must have counted incorrectly, and they did not trust their scoring. Through this direct hands-on experience, we had been forced to face a crucial question: why had we come to believe that big eyes were associated with paranoia in the drawings, when in fact there was no such relationship? The professor reported that naive undergraduate students were just as likely to draw this false conclusion as were more sophisticated PhD students, and even senior psychiatrists and clinical psychologists.

We were stymied and a bit ashamed.

The explanation Professor Chapman gently yet firmly offered to us was that this result involved the occurrence of strong associations—what he termed "strong meaning responses"—especially those connected with emotions.

He further proposed that in the absence of careful record keeping and analytic reasoning, these strong meaning responses would predispose us to develop what he called "illusory correlations" and we would therefore reach erroneous conclusions.

Professor Chapman proposed that when big eyes occurred, for example—especially when they were accompanied by the label *paranoia*—they stood out in our consciousness for both biological and psychological reasons. In humans, as well as primates and other mammals, the eyes appear bigger when one is frightened—the widening of the eyes in fearful situations is well established in research on facial expressions. Not only are our core facial expressions genetically based and hardwired into our nervous systems but we have also witnessed these prototypic facial muscle patterns throughout our lives.

In the research tests, when a picture with big eyes had was labeled "paranoia," Professor Chapman proposed that it evoked a strong meaning response in us. It stood out, and we remembered it.

Since we were not scoring and recording the pictures, we were less likely to remember (1) all the times that big eyes happened not to be associated with the paranoia label and (2) all the times that smaller eyes happened to be associated with the paranoia label. The task of viewing all the pictures was sufficiently complex that we tended to remember the pairings that stood out, in this instance, big eyes with the paranoia label.

Because we were being so distracted by the strong-meaning responses, and since we were not keeping scores and examining the data closely afterward, we would not discover potential true relationship in the data—presuming such relationships were present in the first place. In our class demonstration, since the diagnostic labels were actually randomly associated with the pictures, there were no true correlations in the data to be discovered.

It is factually true that the prototypic facial expression of fear in everyday life is typically associated with big eyes. However, paranoid patients as a general rule do not draw themselves with bigger eyes than do patients with other diagnostic disorders. The fact that we tend to think that they do reflects our tendency, in the absence of formal record keeping, to form illusory correlations and draw illusory conclusions.

One of the fundamental reasons why I became a professional scientist and I learned to incorporate a scientific perspective into my private life—self-science—was to minimize the possibility of fostering illusory correlation experiences and drawing illusory conclusions in any aspect of my life, professional and personal.

As you now understand, I have no wish to be a fool, to fool myself, or to fool you. This is especially the case for something as important as the possible existence of spirits and their potential role in our individual and collective lives.

Self-Science as Protection from Illusion

None of us are immune from forming illusory correlations. My experience is that the best way to minimize this proclivity of the human mind is to consciously employ the evolving analytic tools provided by contemporary science.

This was one of Rene Descartes' greatest realizations and contributions to philosophy, science, and life. He realized the power of employing the methods of science to reduce the likelihood of fooling others and ourselves.

You may be wondering, can science sometimes contribute to producing illusory correlations?

Yes, but only if it is applied in a nonscientific way, and therefore is "bad" science.

Let me explain what I mean here.

If a scientist carefully collects data, but does so selectively—seeking data that supports his hypotheses/beliefs and ignoring data that questions or disconfirms his hypotheses/beliefs—then he increases the probability of forming illusory correlations rather than dispelling them. The key here is not to avoid or dismiss data simply because it fails to support your hypotheses/beliefs.

If this sounds similar to Dr. Carl Sagan's idea of the heart of science in appendix B, you are correct.

When I decided to write *The Sacred Promise*, I did so being fully mindful of the existence of the illusory-correlate effect.

I have concluded that the totality of evidence reported in this book—what is scientifically called proof of concept—is not illusory, and that there is a strong promise that future science will validate and extend the core premise of this book.

Yes, a die-hard disbeliever may conclude that the existence of Spirit is an illusion, and therefore it is foolish to think that Spirit can assist us individually or collectively.

APPENDIX C

However, the disbeliever can only draw this assessment by dismissing much of what has been reported in this book, or by unfairly characterizing what has been presented as evidence of illusory correlation itself.

Since strong disbelievers will likely have much to complain about concerning this book, it is prudent that we address the question of healthy and unhealthy skepticism. I invite you to read about this in appendix D.

APPENDIX D

Healthy and Unhealthy Skepticism about Spirit

The skeptic does not mean him who doubts, but him who investigates or researches, as opposed to him who asserts and thinks that he has found.

—Miguel de Unamuno

Does the collection of innovative, exploratory evidence reported in this book present a compelling case for advocating the creation of a major program of research dedicated to answering this fundamental scientific question: is Spirit real and can Spirit play an important role in our personal and collective lives?

When viewed as a whole, do the pieces form a compelling case to take this work forward?

Have we established clear proof of possibility, if not proof of probability?

Your response to this political question will depend upon whether you are a convinced believer, an unsure agnostic, or an adamant disbeliever.

Readers please be warned—some of the language in this appendix is unpleasant when I discuss the phenomenon of super-skepticism. The reason is, to honor the spirit of us looking honestly and fairly in the mirror, I have attempted to portray superskeptics' style of logic and language clearly and accurately.

The Case to a Convinced Believer

If you already are a believer, the collection of innovative, exploratory evidence reported in this book will seem validating for you, and you may be having feelings of relief as well as of excitement.

For you, the glass was half-full already, and it may seem much fuller now. You probably do not need more research for yourself, at least in terms of belief, but you might be excited about seeing more research conducted if it promised to help you and all of us to improve our capacity to effectively and safely connect with Spirit and benefit from the collaboration.

You might also want more definitive research so you can show your family and friends who were unsure or disbelieving that your core beliefs were ultimately correct. You have probably been teased, criticized, or ostracized for your beliefs, and this is often emotionally painful.

Some of you may already be exploring the role of Spirit in your own lives, and this book may inspire you to explore further.

The Case to an Unsure Agnostic

If you are an unsure agnostic, you may continue to feel baffled, but you see where the totality of the exploratory research is pointing, and your response is probably like mine: if we can get answers one way or the other, let's do it!

Feeling unsure—especially about something as fundamental and big as the spirit-assistance question—is sometimes an unpleasant experience. Because you are not biased against the observations reported in this book, and because the thesis that Spirit may be real and beneficial is not distasteful to you (recalling the oyster metaphor), you can probably see past the inherent limitations of the current state of the science, which is essentially in its infancy, and you can see its great promise.

Since you are open to the possible positive outcomes of this research, you can understand and appreciate my concerted efforts to write this book in as organized yet simultaneously as playful a manner as possible, so that it could be read and understood by the widest possible audience.

You can understand and appreciate that I have purposely restricted presenting excessive details about specific experiments or data analysis procedures so that the forest does not get lost for the trees. You will not critique this book as if it were a formal scientific paper because you realize that this was not the book's intent.

You can comprehend how the three overlapping areas of (1) afterlife research, (2) spirit-assisted healing research, and (3) spirit-guides research together speak strongly to the big questions of "Is Spirit real?" and "Can it play an important role in our individual and collective lives?"

You understand it is the combination of these three areas, with their respective individual questions, that illustrates how the whole is greater than the sum of its parts.

If you are philosophically an orthodox agnostic, as I am, you will be the first to point out that although the work to date does not present a definitive case for the existence of Spirit or spirit-assistance, the totality of the available evidence presents a compelling case for substantially increasing our research efforts. In doing so, scientists will then be able to replicate and extend these intriguing observations and possibilities, and determine if they hold up as predicted.

If they do hold up—and I strongly anticipate that they will—you will have the evidence you require for shifting your position from agnosticism to belief.

And you may even be open to exploring the possibility of testing whether Spirit can play a role in your personal life, and become a self-scientist in the area yourself.

The Case to an Adamant Disbeliever

If you are a disbeliever, and an adamant one to boot, you may feel that the case I have presented is anything but compelling, and you may prefer to describe it as a ridiculous joke.

In fact, to the most ardent ultra-skeptics among you, even the idea of making a case for such work is absurd. (The language used here intentionally reflects the kind of intense and dogmatic style that ardent skeptics have typically employed against this type of work.)

If you are a serious disbeliever, and if you actually have read this book, your feelings may range from disbelief at the audacity of my presenting such weak and flimsy evidence based on flawed experiments and biased anecdotes, to suspicion about my motives, and even to incredulity and disgust toward my colleagues and me.

Some of you may even question the ethics and morals of my current employers—the University of Arizona and Canyon Ranch—for allowing such research to take place at all.

Moreover, you may question my sanity for entertaining the possibility that a deceased person like Susy Smith might wish to continue to speak with me and to do so within the framework of what I call self-science, which you would be quick to label as pseudoscience and/or quack science.

HEALTHY AND UNHEALTHY SKEPTICISM ABOUT SPIRIT

If you are not an ardent disbeliever—instead you are a believer or you are unsure—you may find such statements as these to be unfair. You'll likely interpret them as being untrue.

Since the above hypothetical disbeliever's statements have in reality been said about me, I will be more forthright in my response to them.

Such statements are indeed unfair and are so far from the truth as to border on the pathological.

Of course you realize that since I am writing this book, and am now purposely playing the role of the disbeliever, I have just written all these unfair and untrue things about myself!

Believe it or not, I understand and appreciate the mind of the disbeliever.

I was trained to be highly skeptical. I can readily put myself in his or her shoes and can even play the superskeptic's game—to be bluntly honest, the fact is that for some of them it is both a game as well as a paid profession.

I have come to know a number of professional superskeptics, and I even received one of their infamous yearly awards. In 2001 I learned that I was named as a winner of something called the Pigasus or Flying Pig award (sarcastically adopted from the Pegasus Award), supposedly for being the scientist "who said or did the silliest thing related to the supernatural" for "testing" mediums, "despite pleading by embarrassed friends and colleagues at the U of A." These quotes were taken from the actual statement of the award.

The "award" was partially an attempt at humor; however, the pleading comment, as well as many other statements, were actually gross exaggerations, if not patently false.

If we are to listen to the true disbelievers, *The Sacred Promise* is worse than a forgery; it is actually dangerous because it encourages naive people to believe in the equivalent of Santa Claus or the Easter Bunny. *The Sacred Promise* inspires people to hold on to scientifically debunked myths and superstitions; it is a disgrace to the responsible methods and discoveries of science, and it should be crushed.

If these comments sound extreme and fiery, it is because I am attempting to mimic the ardent disbeliever's language and style as expressed elsewhere.

Though it is not my preference, it is sometimes the case—as experienced firefighters know—that to contain a dangerous fire, you need to respond to it with fire. We will fight fire with fire shortly.

What many disbelievers would have us believe is that they are not disbelievers. Instead they profess, often belligerently, that they are true skeptics. They loudly proclaim they have truth on their side and our best interests at heart.

I once believed that their proclamation was genuine. I have since learned otherwise.

It is essential that we address the skeptic question in this book, because there exists a serious problem in discriminating true skeptics from pseudoskeptics, and genuine explorers from dogmatists, for these two groups of people will view the evidence and the case very differently.

The questioning mind is a remarkable thing, and I believe it should be celebrated. In fact, the questioning mind may be one of our greatest and most important gifts, and ultimately may even be sacred.

Nonetheless, I am reminded of a prudent statement: "Moderation in everything, including moderation," as well as my extension of it to questioning: "Question everything, including the questioning of everything."

Let us briefly look at the light and dark sides of what is called "skepticism," and become crystal clear about when skepticism is true and healthy, and when it is pseudo and becomes dangerously pathological.

What Is True and Healthy Skepticism?

There is such a thing as honest and healthy skepticism—meaning responsible questioning and criticizing—which is essential for distinguishing fact from fiction and understanding from fantasy. Being an orthodox agnostic philosophically, I have a passion for skepticism as an expression of wholesome questioning.

I agree with Einstein, who said, "Imagination is more important than knowledge." However, we should add an important caveat.

Laboratory scientists, as compared with theoreticians like Einstein, have the responsibility to discern the difference between imaginative explorations that are ultimately connected to the real world and fantasies that are not.

Theoreticians focus more on imagination, and researchers focus more on evidence. Though Einstein focused on the former, he deeply cared about the latter as well.

Is Spirit real, or is it a fantasy?

Theories and speculations about the spirit hypothesis—either pro or con—are only meaningful to the extent that they help us discover scientifically whether Spirit is real or not. And while I enjoy the process of imagining and theorizing, it is my nature to do experiments.

Though I love ideas, I adore evidence more.

For me, designing experiments and analyzing data can be an exhilarating experience, especially when the data is surprising.

The challenge of considering all possible explanations without prejudice, and figuring out ways of determining which one (or ones) of them actually account(s) for the greatest amount of the data, is for me as much fun as reading great mysteries.

If someone is going to raise questions and criticisms about scientific research, they should do their homework and know the details of the exploratory investigations as well as the formal experiments and findings. The phrase "not seeing the forest for the trees" is especially important when a veritable forest of innovative exploratory experiments and observations exist, as is the case in this book.

In fact, we cannot accurately evaluate the emerging evidence consistent with the reality of the spirit hypothesis unless we examine all of it and attempt to see the big picture.

We do not want to find ourselves in the shoes of George Seurat's mistress, as portrayed in Steven Sondheim's musical *Sunday in the Park with George*, who saw "all of the parts, but none of the whole."

At the same time, as someone who also appreciates and enjoys trees and their components, including branches, twigs, leaves, trunks, bark, and roots, I try not to lose the trees—and their component parts—for the forest.

In other words, honest and responsible skepticism requires that we (1) not lose the whole for the parts and (2) not lose the parts for the whole.

Honest and responsible skepticism requires that we learn to analyze data as a skilled artist creates a painting. Let me explain.

Some of the time she focuses her attention on individual brush strokes; other times she steps back to see the larger context of what she is trying to paint. One cannot be achieved without the other; it is a dynamic and creative dance between seeing up close and farther back that creates great paintings. Our challenge is to switch between levels, from wholes to parts and back again, and make sure that we are seeing the *whole* picture, which includes seeing the *holes*.

It is also my philosophy, partly owing to my training in clinical psychology and my focus on healthcare, to prefer to treat research questions like people: with kindness, respect, and caring.

Scientists are not necessarily people oriented. It is understandable that we may not be as emotionally and spiritually sensitive as those trained in the healing professions. This becomes critical in afterlife research, where we are dealing with people with feelings and intelligence and who happen to be deceased.

One of the mottoes in our laboratory is "If it is real, it will be revealed; if it is a fake, we'll find the mistake." For open-minded questioners, the journey of discovery is more important than where the journey actually takes us.

If the journey of discovery takes us to Spirit because Spirit is really there, then that is where we will go.

If the journey of discovery takes us away from Spirit because there really is no Spirit, then that is where we will go.

At the present time, the exploratory evidence is pointing to the possible (if not probable) reality of Spirit. The presumption, however, is that you can see the big picture of what this book has presented, and that you are willing to entertain the possibility that not only is the data real but that the implications may be verified in future research.

But what if you can't accept the exploratory data—you are convinced it is impossible—and you find the implications both inconceivable and potentially distasteful? In other words, what if you are an ardent disbeliever?

What you are about to read is somewhat unpleasant and sadly true. Though I would have preferred not to have included a detailed discussion along this line, the fact is that this book has the potential to evoke significant negativity in a group of vocal critics, and for the sake of the work and its integrity, we need to address it more thoroughly.

The discussion will get a bit fiery. We are about to fight fire with fire.

What Is Pseudo- and Pathological Skepticism?

As Warren McCulloch, PhD, said, "Don't bite my finger, look where I'm pointing."

Unfortunately, there are some people who not only wish to "bite my finger" but in some cases would choose to bite off my head as well.

Given their strong biases against the possibility of the spirit hypothesis being true, there is a high probability that they may have a conniption as well as a field day as they read this book.

Rupert Sheldrake, PhD, has created a comprehensive website devoted to exploring the science of healthy and unhealthy skepticism. It is www.skepticalinvestigations.org, and I strongly recommend that you explore it. Sheldrake draws the distinction between inquiry/discovery and dogmatism-doubt. Those who focus on inquiry and discovery are true skeptics; those who maintain dogmatism and denial are said to practice pseudoskepticism. They often show certain characteristics, including:

- Expressing hostility, if not downright meanness, toward certain ideas as well as researchers who investigate them
- Using extreme statements such as "that's impossible" or "it's all fake"
- Ignoring or rejecting important information that does not support their biases
- Searching for the weak or broken twigs, and using them to conclude that there is no forest—in other words, overgeneralizing using tiny or insignificant experimental details or evidence as a means of dismissing the research as a whole
- Attacking the personality and values of the investigator as a means of discrediting the findings
- Making false statements unconsciously or consciously about the research or the investigators
- Engaging in unethical actions to distort or destroy an investigator's laboratory or reputation

Over the years, I have endured repeated instances of the above tactics. A small number of professional skeptics, as well as some virulent members of skeptical organizations, have previously accused me of being:

+ Gullible
+ Sloppy
+ Biased
+ Irrational
+ Dishonest
+ Fraudulent
+ Unethical
+ Crazy

Some of their specific comments have been malicious, patently false, and potentially worthy of legal action. For example, one well-known science journalist and pseudoskeptic once summarily dismissed all of the research in *The Afterlife Experiments* book, as well as my laboratory and me when my name was brought up on a television program.

I quote from the transcript of the television show (at the time this appendix was written, the full transcript was available from CNN at http://transcripts.cnn.com):

> Dr. Gary Schwartz believes in the Tooth Fairy, he believes in UFOs, he believes in levitation, he believes in, as I say, the Tooth Fairy. So he is not a credible scientist.

This pronouncement was made on *Larry King Live* and was probably witnessed by millions of viewers. I was not on the show to defend myself.

Even if I did believe in the Tooth Fairy—which I do not—is this a reason to summarily reject a series of increasingly controlled scientific experiments under blind conditions?

Make note that this supercritic attacked me and my supposed beliefs, and not my experimental evidence.

Let's briefly consider what a sample of pseudoskeptics might say about *The Sacred Promise* and how an objective third person who actually knows the facts might respond to them.

Though some of you might suggest that I am giving such claims more weight by stooping to defend myself, the sad truth is that if I do not illustrate the nature of the arguments and provide sample rebuttals, the pseudoskeptics' fallacious antics go unchallenged and are often presumed to be valid when they are not.

I warn you, what I have written below, which is a fictional portrayal, mimics the style and flavor of pseudoskeptical critiques of our work over the years. At times the language is not kind and the images are not pretty.

Not all pseudoskeptics are so colorful or nasty, but have adopted the style of certain highly visible pseudoskeptical journalists, such as the creator of the Pigasus Award.

Please remember that I am intentionally saying these nasty things about myself to help you understand their antics. But first, there is a little more history.

The He Said–She Said Challenge of Addressing Pseudoskeptics

The problem with addressing pseudoskeptical claims is that it takes significant time and energy to correct them. Moreover, the process of correcting their errors is often experienced as tit for tat, like children squabbling in a sandbox, or sounding like he said, she said in a divorce court.

Here is one sample of a representative pseudoskeptical paragraph containing numerous erroneous and snide comments written by the creator of the Pigasus Award and published on his website in 2001, followed by a brief set of corrections and commentary prepared by me. It expresses the prototypic flavor and style of this pseudoskeptic's criticisms, which I then modeled for the sake of illustration.

Since I know that this gentleman appreciates humor (he was a professional entertainer and can be very funny), I will refer to him as Mr. P. (in honor of his playful award).

Since Schwartz has admitted that he's never done a double-blind experiment, insisting that when he does get around to that mode he will improve it to "triple-blind"— whatever that means!—I will await his implementation of proper controls before making further comment; there is no need to explain something that has not yet been shown to exist. What he has done so far appears to be a series of games and amateur probes, quite without any scientific value.

Corrections and Commentary

First, one of our very first experiments conducted with a medium, which was completed well before the 1999 HBO demonstration experiment, was double-blind and was described in detail in an early chapter in *The Afterlife Experiments* book.

Mr. P. was explicitly told about this double-blind experiment with "proper controls," but he apparently forgot or ignored it and falsely said "when he gets around to it" and "never" instead.

The basic design of a more sophisticated triple-blind experiment was carefully explained to Mr. P., but it may have been too complex for him to comprehend. When Mr. P. writes, "Whatever that means!" he shows that it was he who did not understand the scientific need to improve upon conventional double-blind controls.

When Mr. P. labels our research as a "series of games" and "amateur probes" that were "quite without any scientific value," this illustrates the extreme language, disbelieving bias, and implicit sneering that sadly is all too typical of many pseudoskeptics.

Unfortunately, if a scientist does not respond to egregious false allegations made by pseudoskeptics with appropriate corrections from time to time, the reader may presume that the pseudoskeptics are offering valid criticisms when they are not. Sometimes it is necessary to stand up for the truth, even if it is unpleasant and tiring to do so.

I offer these seven fictional examples of pseudoskeptical criticisms to illustrate how some strongly disbelieving individuals may respond to the position taken in this book: the premise that it is timely and justified to increase our scientific efforts to discover if Spirit is real and can play an important role in our individual and collective lives.

To make this hypothetical he said–she said exercise as impersonal and fair as possible, the fictionalized pseudoskeptics are simply referred to by number, and a created second person provides the corrections and commentary.

Pseudoskeptic #1

This research is all exploratory. Even the few controlled experiments are preliminary. None would be published in mainstream scientific journals. Dr. Schwartz generalizes from flimsy evidence and makes extreme suggestions that go far beyond the "data." There is nothing here.

Corrections and Commentary for Pseudoskeptic #1

This is a cleverly deceptive and unfounded critique.

First, Dr. Schwartz himself emphasizes in the first two chapters that the investigations were exploratory and were proof of principle (proof of concept) and that the majority would not be published in mainstream journals—though some are sufficiently complete to be published in peer-reviewed journals.

He spends significant time explaining to the reader why the wealth of results, though preliminary, deserves serious consideration.

After the pseudoskeptic questioned the significance and implicit credibility of these exploratory investigations, she then overgeneralizes and erroneously infers that the evidence must be "flimsy." This undoubtedly refers to the personal self-science investigations, which by their very nature are meant to suggest future possible laboratory science.

As for the pseudoskeptic's claim that Dr. Schwartz made "extreme" suggestions that went "far beyond the data," the facts clearly indicate that he regularly considered alternative hypotheses, and he reminded the reader that his conclusions were tentative.

It is a typical practice of pseudoskeptics to misrepresent experimental parameters and interpretations in order to draw erroneous negative conclusions.

Pseudoskeptic #2

Talk about imagination. This book is a flight of fancy, a collection of disconnected pseudo-experiments and personal experiences that show us how irrational and crazy Schwartz has become.

Corrections and Commentary for Pseudoskeptic #2

This pseudoskeptic labels the exploratory investigations and experiments as "disconnected." It appears that he did not appreciate or comprehend the intuitive flow of Dr. Schwartz's presentation, often predicated on their real-life venues.

The pseudoskeptic further labels the investigations and experiments as "pseudo," but he does not tell us his criteria. Would he label the replicated biophoton imaging and silicon photomultiplier investigations as pseudo?

In point of fact, these university laboratory investigations employ the basic methods of experimental science.

In addition, what criteria did this pseudoskeptic use to conclude that the author was "irrational and crazy"?

Is it because Dr. Schwartz systematically addresses the controversial questions arising from afterlife research?

Or is it because he is encouraging scientists and the general public to consider that such exploratory investigations and experiments might be revealing a fundamental property of the Universe?

It is neither irrational nor crazy to raise legitimate questions, even if they are controversial, when they can be brought into the laboratory and put to experimental testing.

Pseudoskeptic #3

Yes, Schwartz may be a good storyteller, and even paint a pretty picture with words. But so what? The painting is from his imagination; there is no reality here.

Corrections and Commentary for Pseudoskeptic #3

Pseudoskeptic #3 draws the overarching and erroneous conclusion that there is "no reality here." I think this quote can be aptly rephrased to say, "The reality here is not one I'd like to explore."

The pseudoskeptic covers his reluctance by calling Dr. Schwartz "a good storyteller" and dismissing a large collection of observations, exploratory investigations, and formal experiments as mere imaginative flights.

If pseudoskeptic #3 were genuinely interested in experimental details and numbers, she could have read some of the author's scientific publications on related subjects in peer-reviewed journals.

But this would have refuted her fallacious argument.

Pseudoskeptic #4

The people whom Schwartz works with are at least as weird if not as crazy as he is. How does this woman named Mary, for example, know that she is talking to saints and angels? Schwartz blindly takes these people's experiences at face value and has succumbed to their New Age gibberish. They are pathetic, and so is he.

Corrections and Commentary for Pseudoskeptic #4

Pseudoskeptic #4 makes a set of unsubstantiated claims that the author "blindly" took people's experiences "at face value," and he "succumbed" to their "New Age gibberish."

For pseudoskeptic #4 to make such patently false statements suggests one of the following:

1. He did not read the book.

2. He read it but did not understand or remember what Dr. Schwartz wrote.

3. He is being extreme and overgeneralizing.

4. He is lying.

Dr. Schwartz repeatedly described how he regularly questioned what Mary said and claimed, and he took none of it at face value. Dr. Schwartz tested Mary under a variety of different situational and experimental conditions.

Often pseudoskeptics make unfounded accusations that are based on no information or ignore significant information that would invalidate their arguments.

Pseudoskeptic #5

Schwartz is grieving the loss of his adopted grandmother and is grasping at straws concerning her continued existence. We are forced to take his word on many of the claims. He rarely shows us all the raw data. For all we know, he is making it up, or at least not perceiving it correctly.

Corrections and Commentary for Pseudoskeptic #5

Pseudoskeptic #5 claims that Dr. Schwartz is "grasping at straws" concerning Susy's possible existence.

However, pseudoskeptic #5 fails to mention the complex combination of reasons that actually led Dr. Schwartz to pursue this question with Susy, including the fact:

1. It was her personal wish to "prove that she was still here."

2. She was the author of thirty books in the field of parapsychology and life-after-death.

3. She was a recognized authority on these topics.

Pseudoskeptic #5 is correct that Dr. Schwartz rarely reports all the raw data of a given exploratory investigation or formal experiment, but this doesn't consider the nature of the presentation: a book for a general mass audience.

Dr. Schwartz can refer the skeptic to his many scientific papers and their complete research data.

The statement that Dr. Schwartz might be making it all up implies that his colleagues and research subjects were making it all up, too, a highly unreasonable scenario.

The accusation that it might be all misperception requires that the reader ignore the investigations and experiments where Dr. Schwartz provided actual content and his accurate interpretations of it; the facts clearly do not support this accusation.

Pseudoskeptic #6

Obviously Schwartz is biased. He says he had some personal experiences, even healing experiences. If Schwartz found negative evidence, he probably won't see it or report it or understand it. Schwartz is deceiving himself and the reader about his claims of putting all the possible explanations on the table.

Corrections and Commentary for Pseudoskeptic #6

This comment illustrates how little pseudoskeptic #6 knows of the author's almost forty years in science and academia.

He makes the accusation that if Dr. Schwartz found negative evidence, "he probably won't see it or report it or understand it." The facts clearly indicate otherwise.

Dr. Schwartz and his colleagues have published numerous negative studies, including a report of a woman claiming psychic abilities whom they discovered in the laboratory was cheating, as well as a detailed analysis demonstrating how so-called orbs are almost always caused by inexpensive lenses on cameras and light reflections.

In the book itself, the author describes numerous negative observations and apparent experimental failures.

And in each of Dr. Schwartz's books, including this one, he devotes pages to carefully examining alternative hypotheses.

Pseudoskeptic #7

Skeptics have already pointed out the fatal flaws in Schwartz's previous research. Now he presents a few preliminary experiments and anecdotes that happened in some of his experiments and personal life, and we are supposed to take this seriously? Who does Schwartz think he is?

Corrections and Commentary for Pseudoskeptic #7

A common technique of pseudoskeptics is to make claims that certain imperfections or limitations in a given investigation or experiment are "fatal."

Pseudoskeptic #7 takes this one step further by claiming that the author's previous research was fatally flawed, and the new book's research is much weaker.

The accusation of fatal flaws is serious, and deserves special attention. To place such a claim in context, let's review a serious flaw reported on the internet and even in the *Skeptical Inquirer* magazine and consider whether it deserves the extreme label of fatal.

A well-known pseudoskeptic, Mr. P. noticed that the screen separating the mediums and the sitters in the author's HBO demonstration experiment had a tiny slit just big enough to look through.

Mr. P.'s pseudoskeptical colleagues jumped on this observation, claiming that it showed how naive and sloppy the author and his colleagues were, and how this might explain his amazing results.

In announcing Dr. Schwartz's receipt of a 2001 Flying Pig Award, Mr. P. wrote:

> As a single example of his poor control of the "experiments" in "carefully isolating" a subject from a psychic during the tests—so that the psychic would have no information about the subject—he failed to notice the psychic clearly peeking into the adjoining area. When this was pointed out to him, he shrugged it off as an unimportant factor. So much for scientific rigor.

What Mr. P. and his colleagues failed to mention was that during the readings the mediums were not facing the tiny crack; they could only see the sliver when they were first sitting down and looking at the screen.

The video recordings clearly show that during the readings, the mediums were facing the video camera and never looked sideways to potentially peer through the crack. This was obvious in watching the HBO documentary.

Do you believe that it is reasonable to conclude that a quick glance made once, at the beginning of a reading, could plausibly explain the greater than 80 percent overall accuracy scores that the mediums obtained for the complete readings that typically took ten to fifteen minutes?

Paraphrasing pseudoskeptic #7, who does this pseudoskeptic think he is?

Mr. P. and his pseudoskeptical colleagues also failed to mention the results of the EEG and EKG readings in the HBO experiment, which indicated that when the mediums were doing their readings, they showed less synchronous EKG and EEG with the sitter.

In other words, during the readings, the mediums appeared less connected physiologically to the sitter, which does not support Mr. P.'s interpretation that the mediums were actively looking through the slit for visual cues.

Mr. P. and his pseudoskeptical colleagues also failed to mention the results of other investigations and experiments reported in the book that used slit-less screens and long-distance phone readings, probably because these results were positive and did not support their extremely flawed argument.

When reading critiques written by pseudoskeptics, it is important to carefully evaluate their use of extreme and overgeneralized statements; those words are often a sign that something is not right.

I had originally considered including more examples, but the required he said–she said format becomes tiring for both the reader and me. Hopefully the prototypic examples provided make it possible for you to better appreciate some of the general strategies of pseudoskepticism in action.

APPENDIX D

Do I welcome serious questions, critiques, and feedback? Absolutely, but with the following proviso:

If the people offering the critiques have not done their homework, have misrepresented the facts, or have presented their misinformed critiques in an unfriendly or mean manner, then they are wasting his time and mine.

My personal focus is on discovery, not dogma; my devotion is to true skeptical inquiry and questioning, not to pseudoskeptical debunking.

Having said this, I must confess that my heart goes out to people when they are not aware of the fact that they are being dogmatic and pseudoskeptical.

My students and I have published more than twenty research papers in mainstream journals and academic books on the psychophysiology of self-deception and repression. I know what it is like when someone attributes to others what they cannot see in themselves.

I have seen it even in psychologists and psychiatrists.

Advanced academic training, and even psychotherapy, does not necessarily guarantee that one is accurate in one's self-awareness.

It requires regular self-monitoring, nondefensiveness, and a genuine and persistent desire to know oneself. It requires the courage to truly look at oneself in the mirror, face that person, and have the inspiration to change.

This is why I continually question myself and others about our reasoning, sanity, and integrity. I am aware of the potential risk of self-deception and repression, and we try to stay mindful of it.

If you ask someone, "Are you angry?" and he replies, in a sharp and angry tone, "No, I'm not angry!" the disconnection between what he is saying and what he is expressing nonverbally is one indication that a person's self-awareness may be impaired.

Pseudoskeptics are not usually appreciative when someone is skeptical about their beliefs. Pseudoskeptics often become defensive and angry when their beliefs are challenged, and this can apply as much to distinguished academics as it does to people with less than a high school education.

Healthy skeptics, on the other hand, enjoy other healthy skeptics' questioning and challenging what they are doing, and they invite friendly critical feedback.

While I celebrate healthy questioning and skepticism, I deplore unhealthy dogmatism and cynicism.

And yes, I occasionally get angry and defensive at times. What evokes my ire is when a person makes uninformed, biased, and malicious accusations that disrespect honest inquiry and genuine discovery.

To reiterate, *The Sacred Promise* may be wrong in some of its details, and my colleagues and I are fully mindful of this possibility.

However, the emerging experimental evidence, obtained in the university laboratory as well as the laboratory of real life, suggests the serious possibility that there may be a beautiful baby in the bathtub here, and it would be criminal (some might say murder) to throw the baby out with the bathwater.

Celebrating the Opportunity of Addressing
The Sacred Promise Hypothesis

As mentioned previously, Dr. Carl Sagan is one of my heroes. His was a creative and visionary mind, committed to science and the possibility that the Universe was grander than most of us currently imagine.

Dr. Sagan said something that soothed my mind as it touched my heart, which I quoted at the beginning of the appendices. He said:

"When Kepler found his long-cherished belief did not agree with the most precise observation, he accepted the uncomfortable fact. He preferred the hard truth to his dearest illusions; that is the heart of science."

APPENDIX D

The ultimate hope for humanity and this planet is that all of us, in principle, have the potential to learn to accept uncomfortable facts and follow the hard truth rather than our dearest illusions. This is the heart of science; it is also our greatest challenge as well as our finest capability.

We can see beyond illusions, go beyond our limited senses, and as Marcel Proust said, see with new eyes.

It is an illusion that the earth is flat. It just looks that way with our limited vision when we are on the surface of the earth.

It is an illusion that the sun revolves around the earth. It just looks that way with our limited vision when we are stationary on the earth.

It is an illusion that objects are solid. It just looks that way with our limited vision when we see physical objects.

It is an illusion that invisible space is empty. It just looks that way with our limited vision when we process frequencies of light only using the retinal cells of our eyes.

Just because we experience illusions with our limited senses does not mean we are unable to go beyond them. The history of science provides repeated, and I would say definitive, evidence indicating that we can change our minds and go beyond what we once believed, as new evidence appears and awakens us.

What is even more remarkable to me is that we all have the innate capability to learn this meta-lesson, this lesson of lessons. We have the potential to see beyond our biological limitations, and we can evolve and transform our consciousness accordingly.

Though we may love certain ideas and be frightened of others, this does not mean that we are unable to let go of long-cherished beliefs and adopt new ones that may seem at first to be foreign, uncertain, gigantic, and even beyond our current imagination.

The history of science shows us that our minds can do this, and our hearts can catch up. As Dave Palmer wrote, and Carole King sang, "I can see you've got a change in mind but what we need is a change of heart."

What is precious about *The Sacred Promise* hypothesis is that if it is true, then this bigger-than-life vision is full of hope, opportunity, adventure, and discovery. *The Sacred Promise* hypothesis gives new meaning and purpose to this life and life beyond life.

If numerous mediums are correct—and I underscore if—then Carl Sagan has changed his mind about life after death and a larger spiritual reality. So too has Harry Houdini (as discussed in chapter 14).

Only a week before I was writing the first draft of this appendix, two mediums again claimed that Einstein wants to speak with me. More important, he wants to speak with all of us.

And just this morning, as I was working with the copyedited draft of this appendix, a distinguished professor at Tel Aviv University in Israel and her highly skilled intuitive called me, revealing remarkable evidence indicating that the intuitive could bring forth accurate physics formulas from Einstein and other deceased luminary physicists, which can be scientifically verified.

Are we willing to listen to Einstein, Sagan, Houdini, and countless other wise and caring deceased people, if they are still here? Are we willing to listen to Sophia, Michael, Gabriel, and countless other wise and caring spirit guides and angels, if they are here?

Are we willing to listen to the Great Spirit, the Source, the Sacred, if She/He/It is here?

I would hope that if future science reveals that they are here, that we will be able to honor Sagan's vision and wisdom. That we will be able to let go of the illusion that all there is is the physical world, and sooner rather than later embrace the truth that they are here, with us and for us.

This is the heart of science. And also of *The Sacred Promise*.

RECOMMENDED READING

This is an abbreviated reading list addressing science, consciousness, spirit, healing, and the sacred. These books provide a background and inspiration for *The Sacred Promise*. All twenty-one titles are available at www.amazon.com and other major booksellers.

Science and Survival of Consciousness

D. Blum. *Ghost Hunters: William James and the Search for Scientific Proof of Life After Death* (Penguin Press, 2006).

S. E. Braude. *Immortal Remains: The Evidence for Life After Death* (Rowman & Littlefield Publishers, 2003).

D. Fontana. *Life Beyond Death: What Should We Expect?* (Watkins Publishing, 2009).

S. Horn. *Unbelievable: Investigations into Ghosts, Poltergeists, Telepathy, and Other Unseen Phenomena, from the Duke Parapsychology Laboratory* (Ecco, 2009).

G. E. Schwartz, with W. L. Simon. *The Afterlife Experiments: Breakthrough Scientific Evidence for Life After Death* (Atria Books, 2003).

G. E. Schwartz, with W. L. Simon. *The Truth about* Medium: *Extraordinary Experiments with the Real Allison DuBois of NBC's* Medium *and Other Remarkable Psychics* (Hampton Roads Publishing, 2005).

P. van Lommel. *Consciousness Beyond Life: The Science of the Near-Death Experience* (HarperOne, 2010).

Science and the Paranormal

L. Dossey. *The Power of Premonitions: How Knowing the Future Can Shape Our Lives* (Dutton, 2009).

D. H. Powell. *The ESP Enigma: The Scientific Case for Psychic Phenomena* (Walker & Company, 2009).

D. Radin. *Entangled Minds: Extrasensory Experiences in a Quantum Reality* (Paraview Pocket Books, 2006).

C. T. Tart. *The End of Materialism: How Evidence of the Paranormal Is Bringing Science and Spirit Together* (New Harbinger Publications, 2009).

R. Sheldrake. *Morphic Resonance: The Nature of Formative Causation* (Park Street Press, 2009).

RECOMMENDED READING

Science, Energy, and Healing

D. J. Benor. *Consciousness, Bioenergy and Healing: Self-Healing and Energy Medicine for the 21st Century (Healing Research, vol. 2, Professional Edition)* (Wholistic Healing Publications, 2004).

L. Dossey. *Healing Words: The Power of Prayer and the Practice of Medicine* (HarperOne, 1995).

R. Gerber. *Vibrational Medicine for the 21st Century: A Complete Guide to Energy Healing and Spiritual Transformation* (William Morrow, 2001).

J. Oschman. *Energy Medicine in Therapeutics and Human Performance* (Butterworth-Heinemann Publishing, 2003).

G. E. Schwartz, with W. L. Simon. *The Energy Healing Experiments: Science Reveals Our Natural Power to Heal* (Atria Books, 2008).

Science, Spirituality, and God—Skeptical and Possible

A. Goswami. *God Is Not Dead: What Quantum Physics Tells Us about Our Origins and How We Should Live* (Hampton Roads Publishing, 2008).

B. B. Hagerty. *Fingerprints of God: The Search for the Science of Spirituality* (Riverhead Books, 2009).

G. E. Schwartz, with W. L. Simon. *The G.O.D. Experiments: How Science Is Discovering God in Everything, Including Us* (Atria Books, 2007).

V. J. Stenger. *Quantum Gods: Creation, Chaos, and the Search for Cosmic Consciousness* (Prometheus Books, 2009).

INDEX

INDEX

INDEX

INDEX

INDEX

INDEX

INDEX

INDEX